T0134873

Wireless Networks

Series editor

Xuemin (Sherman) Shen
University of Waterloo, Waterloo, Ontario, Canada

More information about this series at http://www.springer.com/series/14180

Xiang Cheng • Luoyang Fang • Liuqing Yang
Shuguang Cui

Mobile Big Data

 Springer

Xiang Cheng
State Key Laboratory of Advanced
Optical Communication Systems
and Networks
School of Electronics Engineering
and Computing Science
Peking University
Beijing, China

Liuqing Yang
Department of Electrical &
Computing Engineering
Colorado State University
Fort Collins, CO, USA

Luoyang Fang
Department of Electrical &
Computing Engineering
Colorado State University
Fort Collins, CO, USA

Shuguang Cui
Department of Electrical
and Computer Engineering
University of California - Davis
Davis, CA, USA

ISSN 2366-1186   ISSN 2366-1445 (electronic)
Wireless Networks
ISBN 978-3-030-07145-5   ISBN 978-3-319-96116-3 (eBook)
https://doi.org/10.1007/978-3-319-96116-3

This Springer imprint is published by the registered company Springer Nature Switzerland AG.
The registered company address is: Gewerbestrasse 11, 6330 Cham, Switzerland

Preface

Since the appearance of the first commercially automated cellular network launched by Nippon Telegraph and Telephone (NTT) in 1979, mobile network technology has become a necessity during the past four decades of amazingly rapid development. In 2009, the Long-Term Evolution (LTE) network (the most popular fourth-generation standard) was first deployed in Oslo, Norway, and Stockholm, Sweden. Since then, mobile phones (smart phones) have successfully penetrated nearly every aspect of human life, due to flourishing mobile applications and services. At the same time, massive data generated by mobile devices during mobile network operations and at backend servers, termed as mobile big data, has attracted significant attention from various research communities and industries. However, large-scale collection and analysis on mobile big data only became possible in the past decade, due to the highly demanding computing and transmission capability in dealing with such tremendous volume of mobile data, which are vastly lacking until recently. One of the most distinct characteristics of mobile big data is its spatiotemporal feature, which provides the timestamp and location information of a certain user on every record of the mobile big data. As a result, the mobility of human being is first studied based on the highly informative mobile big data in the literature. Behavior patterns revealed by the mobile big data can facilitate many novel data-driven applications spanning subjects from personalized location-based recommendation and pervasive health computing to aggregated public services including urban planning and network management. However, the personal information inherently contained in mobile big data may lead to a privacy concern.

This monograph provides a comprehensive picture regarding the life cycle of mobile big data, starting from the data source and collection, transmission and computing all the way to applications. In Chap. 1, the mobile big data is introduced and its characteristics are summarized. In Chap. 2, mobile data sources are overviewed in two categories, namely, the app level and the network level, and the data collection in the mobile network is extensively explained, together with the description of the LTE network architecture. In Chap. 3, the supporting infrastructure on communications and networks for mobile big data transmission is surveyed, in which the challenges brought by mobile big data are also described.

In Chap. 4, the computing architecture and paradigm are introduced for large-scale data processing and analytics, in terms of the distributed computing hardware and the map-reduce-based software. In Chap. 5, the big picture on mobile data-driven applications are sketched, together with a brief introduction of machine learning and data mining techniques. In addition, the user profiling and modeling are presented in detail, which provide a foundation for many personalized data-driven applications. In Chaps. 6 and 7, two spatiotemporal analysis cases on mobile big data are presented based on a signaling dataset collected by a mobile network operator in urban areas. Chapter 6 focuses on the aggregated spatiotemporal learning in terms of cell-wise demand forecasting for predictive network management, whereas Chap. 7 spotlights on the individual spatiotemporal analysis from the perspective of privacy attacks. These two chapters are expected to give vivid examples of mobile big data and its related data analysis and mining.

The potential readers of this monograph are researchers, graduated students, and professors relevant to this field. This monograph also provides the state of the art on mobile big data for people outside this field and aspires to trigger new directions and research ideas of this interdisciplinary field.

We would like to thank Dr. Haonan Wang, Dr. Rongqing Zhang, and Dr. Dexin Wang for their inspiring discussions on the research work presented in this monograph. Finally, we would like to thank the continued support from the National Natural Science Foundation of China under Grants 61622101 and 61571020 and the National Science Foundation under Grants DMS-1521746 and DMS-1737795.

Beijing, China   Xiang Cheng
Fort Collins, CO, USA   Luoyang Fang
Fort Collins, CO, USA   Liuqing Yang
Davis, CA, USA   Shuguang Cui

Contents

Acronyms

3GPP	The 3rd Generation Partnership Project
5V	Volume, Velocity, Variety, Veracity, Value
ARIMA	Auto Regression Integrated Moving Average
CDR	Call Detail Records
CN	Core Network
CNN	Convolutional Neural Network
CPT	Control-Plane Traffic
CS Core	Circuit Switched Core
EMM	EPS Mobility Management
EPS	Evolved Packet System
GCN	Graph Convolutional Network
GPS	Global Positioning System
GRN	Gated Recurrent Network
IMEI	International Mobile Equipment Identity
LTE	Long-Term Evolution
MBD	Mobile Big Data
MDC	Mobile Data Challenge
MinBM	Minimum-Cost Bipartite Matching
MLDM	Machine Learning and Data Mining
MME	Mobility Management Entity
OTT	Over The Top
PACF	Partial Autocorrelation Function
PCEF	Policy Control Enforcement Function
PCRF	Policy and Charging Rule Function
PGW	Packet Data Network Gateway
PS Core	Packet Switched Core
RAN	Radio Access Network
RCC	Radio Resource Control
RCN	Radio Control Network
RDD	Resilient Distributed Dataset
RMR	Radio Measurement Report

SDN	Software Defined Networking
SGW	Serving Gateway
SLAM	Simultaneous Localization and Mapping
SSID	Service Set Identifier
TA	Tracking Area
UE	User Equipment
UPT	User-Plane Traffic

Chapter 1
Mobile Big Data

1.1 Overview of Mobile Big data

The smart phone evolution in the past decade has accelerated the proliferation of mobile Internet and spurred a new wave of mobile applications on smart phones. In particular, GPS is becoming part of the default configuration of any smart mobile devices, rendering location information readily available. Even in the lack of exact location information when GPS is not enabled, the coarse location can still be inferred from the network-level data. The location information alone can already enable a great variety of applications to provide personalized services (context-aware recommendation, next location prediction based traffic time estimation, etc.) and to assist public service planning (e.g., traffic flow analysis, transportation management, city zone recognition, etc.). As smart phones are equipped with a variety of sensors, personal behaviors can be further learned and monitored. In addition, mobile operators can also collect a huge amount of data to monitor the technical and transactional aspects of their networks. It has been recently recognized that such data, known as mobile big data, could well be an under-exploited gold mine for almost all societal sectors.

In the past, non-structured data fragments are usually considered as useless byproducts merely to facilitate the proper flow of structured data. Nowadays, the purpose of big data processing is to piece together such data fragments so as to gain insights on user behaviors, and to reveal underlying routines that may potentially lead to much more informed decisions. Drastically differing from the traditional practice where services determine and define the data, in the big data era, data is becoming a proactive entity that may drive and even create new services.

Compared with the so termed 5V characteristics of generic big data, namely volume, variety, velocity, veracity and value, mobile big data is distinct in its unique multi-dimensional, personalized, multi-sensory, and real-time features [1]. Recent research on mobile big data processing has shown its great potential for diverse purposes ranging from improving traffic management, enabling personal

© Springer International Publishing AG, part of Springer Nature 2018
X. Cheng et al., *Mobile Big Data*, Wireless Networks,
https://doi.org/10.1007/978-3-319-96116-3_1

and contextual services, to enhance public security, etc. For instance, data driven activity recognition is essential for healthcare applications [2]; the usage pattern of smart phones could be utilized to learn the mental status of users [3]; and the mobile data can provide critical information to facilitate the resource optimization in communications networks (e.g., enhancing paging efficiency, provisioning future data rate, predicting resource needs, etc.).

The unique value of mobile big data comes from its ubiquity and context-richness. It has been evident that mobile Internet not only offers traditional services running on the fixed Internet, but also enables a broad range of new applications that allow the Internet to immerse into almost every aspect of our modernizing society. In fact, the mobile Internet traffic carries a much richer context, which pinpoints the time, location, activity, social relationship, and surrounding environment of mobile users. Consequently, mobile big data research has a multi-disciplinary nature that demands diversified knowledge from mobile communications and signal processing to machine learning and data mining. The research field of mobile big data has been booming quickly in recent years, but is somewhat fragmented. This monograph aspires to provide an integrated picture of this emerging field to bridge multiple disciplines and hopefully, to inspire more coherent future research activities. In addition, this monograph also provides mobile big data driven case study to exemplify details of mobile dataset and its related applications. Before digging into the life cycle of mobile big data, we first review the distinct characteristics of the mobile big data.

1.2 Characteristics

As the mobile devices (e.g., smart phones, wearable devices) have become the center of almost everyone's daily life, mining the sheer volume of data from mobile devices has attracted great interests from various research communities, such as data mining, statistics, communications, machine learning, sociology, geography, and so on. This is mainly due to the rich characteristics of mobile big data.

1.2.1 "5V" Features

Mobile big data first inherits the "5V" features of generic big data [4], namely volume, velocity, variety, veracity, and value. Though the concept of big data is not precisely defined, its ubiquitous features are well recognized, rendering big data quite different from some simple massive data. The definition of the first "3V" characteristics (volume, velocity, variety) could be dated back to the report by Laney in 2001 [5] and the remaining "2V"s were emphasized in more recent work [6, 7], which are summarized below in the context of mobile big data.

Fig. 1.1 Distinct
characteristics of mobile big
data

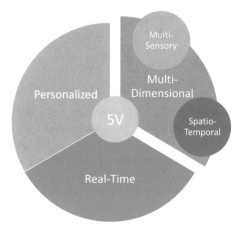

- **Volume.** The *volume* of big data refers to the tremendous size of the data. In the context of mobile data, it is predicted that the mobile data traffic will exceed 15 exabytes per month by 2018 [4].
- **Velocity.** The *velocity* of big data indicates the rapid data generation and streaming. The high penetration of smart devices nowadays, e.g., smart phones, wearable devices, etc., will generate and stream sensed data at an unprecedented speed to facilitate context-aware and personalized applications.
- **Variety.** The *variety* indicates the complexity of mobile big data, which comes from the great heterogeneity in the data types, e.g., multi-sensory data, audio and video footages, etc.
- **Veracity.** The *veracity* suggests the quality of different sources of big data may be inconsistent [6] even in the same domain. Therefore, the data may be noisy, inaccurate and redundant, which should be first cleaned and preprocessed before analysis.
- **Value.** The *value* of big data was first discussed by Gantz and Reinsel [7], who outlined that the big data technologies hinge upon the economical *value* extraction from the massive volume, high velocity and wide variety of data, with the capability of data analysis and knowledge discovery.

Besides the 5V characteristics as for traditional big data, mobile big data also exhibit some distinct features, which will be introduced in the following subsections (Fig. 1.1).

1.2.2 Multi-Dimensional

The multi-dimensional feature is naturally inherent in mobile big data, as it is generated by multiple sensors and tagged with time and geolocation information at varying granularities. Indeed, the *spatio-temporal* and *multi-sensory* features are clearly noticeable in mobile big data. Note that the correlation across dimensions makes this feature distinct from the "variety" feature.

Spatiotemporal

Almost every entry in mobile big data is tagged with a time stamp and certain geolocation information, which enables a great number of new applications. In fact, almost every smart phone is equipped with a GPS receiver, which provides accurate outdoor location information with raw data containing the latitude and longitude. Even when the location service of a smart phone based on GPS is not enabled or not reliable (e.g. when indoor), different granularities of location information can be inferred by other data entries, e.g., service set identifier (SSID) of WiFi access points, cell ID in call detail records (CDR) [8], WiFi signal strength [9–11], and even IP addresses [12–14].

In particular, the CDR data records the time stamps and approximate location information for all calls and text messages of each user, which are automatically generated by the telecommunication systems. However, the CDR cannot provide the location information when the user is not active. That is, no location information is available between two call records. In [8], Ficek et al. proposed a probabilistic model that estimates user locations between their consecutive communication events (calls or text messages), in order to obtain finer trajectories of users from the network cell transition information in CDR.

It is a common consensus that GPS cannot be used indoor. In addition, the location information directly obtained from cell IDs is not sufficiently accurate for certain mobile applications, e.g., location-aware precise mobile advertising. In the literature, localization in indoor scenarios can be achieved by exploiting received WiFi signal strengths. The unpredictability of signal propagation through indoor environments is a major challenge in localization based on WiFi signal strength. Ferris et al. in [10] aimed to build a position-conditioned likelihood model for signal strength distributions based on Gaussian process latent variable models, from which the accurate location information can be learned by using simultaneous localization and mapping (SLAM) techniques without any location labels in the training data. In [11], Huang et al. improved the computational complexity of the method proposed in [10] from $O(N^3)$ to $O(N^2)$ using GraphSLAM, and relaxed several constraints from [10], e.g., limited predefined shapes (narrow and straight hallways). The accuracy of indoor localization in [11] was claimed to be between 1.75 and 2.18 m over an area of $600\,\mathrm{m}^2$.

When the location service is not enabled or when users are not willing to share their location information due to privacy concerns even in the outdoor scenarios, the user location information to some degree could be still learned from the available mobile big data to facilitate mobile applications while protecting user privacy. In [14], Long et al. proposed an approach to infer the user locations from the hashed user IP addresses at the census block group (CBG) level, where CBG is a geographical unit defined by the United States Census Bureau (USCB) and typically has a population of 600–3000.

In addition, the location information is often used to facilitate various recommendation services. However, the raw location information, such as coordinates (longitude and latitude) from GPS receivers, cell IDs from CDRs, or even the indoor location estimated from WiFi signal strength, is meaningless for certain mobile applications (e.g., recommendation services, mobile advertising, etc.), if it is not mapped correctly to what can be understood by human beings. Therefore, tagging the location semantically is critical for many mobile applications. However, it is also challenging, especially when it comes to the extremely dense urban areas, due to the great amounts of location data [15] and the inadequate accuracy of civilian GPS [16]. In [17], Goncalves et al. built a crowdsourcing framework termed as *Game of Words* to interact with users for their personalized semantic tagging of locations. The *Game of Words* identifies, filters, and ranks keywords, by which many users can characterize a location, such that the semantic location tagging could be adapted to dynamic changes of a location without degradation due to noises and biases as with the single-source data.

Multi-Sensory

Almost all smart phones nowadays are equipped with a rich set of embedded sensors [4], e.g., accelerometer, thermometer, compass, gyroscope, GPS signal receiver, ambient light sensor, etc. Such embedded sensors can provide a tremendous volume of data. For example, 1 h of simple personal monitoring (e.g. ECG, HR, accelerometer data, etc.) generates about 14 MB of data [18]. With these embedded sensors, context sensing can be performed to facilitate context-aware applications. However, context sensing requires multiple sensors to provide correlated multi-dimensional data simultaneously such that the sensing result could be more accurate. In other words, a single sensor may be of little use semantically in depicting the context of device holders. With smart fusion of data from multiple sensors, more data-driven mobile applications, such as pervasive health computing, activity recognition, context-aware services and so on, could be facilitated by smart devices.

In addition, with the built-in connectivity, smart phones often serve as sensor hubs for wearable sensors [18], e.g., ECG sensors, pedometers, etc. Though the high dimensional data from multiple sensors provides vast possibilities and great potentials for mobile applications, it also inherits some drawbacks from typical sensor data, e.g., incomplete dataset and outliers, due to random sensor failures. This leads to some interesting but challenging steaming big data problems [19, 20].

1.2.3 Real-Time

A typical requirement of analytics over mobile big data is that the location-based and highly personalized information and services need to be delivered to mobile users nearly in real time [21]. For example, the dynamic processing of mobile big data adapted to the context of the device holder requires the environmental sensing data (e.g., temperature, humidity, etc.) almost simultaneously. On the other hand, smart phones highly penetrate the modern lives of users and can serve as a real-time interaction platform between users and applications. That is, mobile big data record the real-time user preferences given different contexts and scenarios. Therefore, the mobile-big-data-driven applications should respond to user requests in near real-time to ensure the quality of experience, especially in certain time-critical applications such as mobile health.

However, the resource limitation of mobile devices in terms of computation, storage, and battery could hardly meet the intensive demands on massive mobile data processing. In this aspect, the mobile cloud computing schemes may provide a solution, which allocate intensive computing demands between the mobile devices and cloud computing devices [4].

1.2.4 Privacy Sensitive

Mobile data directly collected from user devices or mobile networks (e.g., gateways, base stations) contains user identities. Besides the identity information, the mobile data itself is usually highly personalized and linked to user locations and contexts. In fact, the time-stamped geolocation information records the trajectories of users, which exposes their fundamental privacy. For example, the most visited location of a user at night based on GPS is very likely the physical address of the user. However, from the perspective of mobile big data mining, the privacy-sensitive information are inevitably demanded for precisely personalized mobile applications.

Generally, privacy protection is highly concerned in any personal-data-related services and applications. *k-anonymity* is a typical metric to evaluate the effectiveness of privacy preservation [22], which requires any record in a database to be indistinguishable to at least $k - 1$ other records in the database. The most common anonymization technique is to replace critical identifiers (e.g., phone number, IMEI, etc.) with random pseudo identifiers. However, such identifier anonymization fails for the mobile data with which the subscriber spatiotemporal behavior is recorded, due to the uniqueness of human mobile trajectories. That is, the indistinguishableness is difficult to achieve even in a dataset with a large number of users.

In [23], Zang and Bolot studied a large-scale nationwide dataset with more than 30 billion call records corresponding to 25 million users with different spatial granularities (i.e., cell sector, cell, zip code, city, state). The spatiotemporal footprint of each user is represented by the N most visited places within a pre-defined time period (e.g. day, week, month, etc.), based on which the privacy leakage risk could be evaluated. The authors concluded that the spatiotemporal data sharing or publishing that is only anonymized by pseudo identifiers leads to a severe privacy leakage risk. The potential privacy-preserving solution is to at least coarsen the temporal resolution, which restricts the accuracy of extracting N most visited locations from the dataset. However, such temporal resolution reduction will large reduce the utility of the data. In addition, it is concluded in [24] that spatiotemporal resolution curtailments may not be effective as expected, based on a human mobility study with 15-month mobile data and 1.5 million people in a country. That is, the uniqueness reduction is magnitudes of order slower than the resolution coarsening. Therefore, a generalized scheme on the spatiotemporal privacy preserving based on k-anonymity was proposed in [25].

The user identification (or user reconciliation) is another critical problem in privacy protection, which is to link the spatiotemporal records generated by the same user in two datasets of the same domain [26] or two datasets of different domains [27]. The user identification is closely related to "de-anonymization" attacks. A typical example is the Netflix prize task that is aimed to de-anonymize user identities by public user reviews [28]. In [29], De Mulder et al. studied the user identification based on the location update dataset from GSM networks, which records the phone's network location with geographical information periodically. The mobility Markovian model of each user is constructed based on their spatiotemporal history, including the cell visiting transition probability matrix and cell visiting stationary probability. The user identification is formulated as the heuristic comparison of transition probability matrix and stationary probability between any pair of users in the dataset or searching the user with maximum probability belief for a given observed location update sequence of a specific user based on their transition probability matrices. However, such Markovian model requires the dataset with subscribers' transitions among cells to be recorded, whereas such data is not widely adopted or collected by mobile network operators.

In [26, 27], user identification is formulated as the minimum (maximum) cost bipartite matching with two sets of vertices representing users in two datasets, respectively, where the edge weight is obtained by the distance (similarity) measure between any pair of nodes in the bipartite graph. In [26], Naini et al. suppress the temporal information of users' spatiotemporal trajectories and represent the user fingerprint as the histogram of visited location for a given time length, where the histogram can be viewed as the visiting frequency of each subscriber over each location points. The distance between two histograms is calculated by the Jensen-Shannon divergence. Instead of temporal information suppression, Riederer et al. in [27] models the number of spatiotemporal appearances of a given spatial and temporal bins by Poisson process for each dataset, based on which the similarity scores could be generated. The task of [27] is to identify the user of two datasets from different domains during the same time period.

Therefore, the management and encryption of privacy-sensitive data should be well investigated [30]. In the collection campaign of MDC [31], user privacy was heavily emphasized and protected by careful data collection design. In particular, MDC explicitly guarantees that the data is completely owned by the participants and each individual has the full control rights of their data [32, 33], such as data accessing, data deletion, etc. Also, the identity of users, phone numbers, identifiers of WiFi and Bluetooth nodes are hashed as pseudonyms and the accuracy of location information is mapped to different levels for both privacy protection and data usability. In addition, the data access management for differently authorized privileges should be well designed to regulate the data exposure.

In addition, the trend of mobile big data analytics is not just for analyzing the past or understanding the present, but also for predicting the future [34], which will provide predictive personal services (e.g., smart context-aware personalized services). Therefore, not only the raw data collected are privacy-sensitive, but also will the results mined from mobile big data reveal the daily personal life patterns of users. Therefore, both the data itself and its analysis results should be carefully protected. Otherwise, the availability of data may be in turn jeopardized, for people might end up unwilling to share their data [31].

The semantic extraction of location information could be used to help protect user privacy, as users have options to share their location information through different levels rather than sharing the exact GPS coordinates, e.g., through the levels of city, district, etc. Furthermore, the obfuscation-based techniques may be used to disguise the actual position by providing less accurate or even faked location information [35]. However, if the region level is too coarse, it will jeopardize its usability in mobile applications. In addition, the obfuscation techniques may not be able to protect the privacy of a user, as adversaries may infer the actual location of a user based on their background information. To address this, the location region information can be transformed to different levels, which are carefully designed such that the privacy-sensitive location information may be cloaked without losing too much accuracy. In [35, 36], Damiani et al. proposed a privacy-preserving obfuscation environment (PROBE) framework to personalize the protection of sensitive semantic location, based on the privacy profiles generated by users against the privacy attacks of adversaries.

Summary

Mobile big data inherit some traditional features from generic big data but also have several distinct addons. Its multi-dimensional nature from multiple sensors tagged with fine-grained time stamps and geolocation markers provide fuels to accelerate many personalized precise mobile applications. On the other hand, the real-time response requirement of mobile big data applications and privacy-sensitive data management itself will post a great challenge to system design.

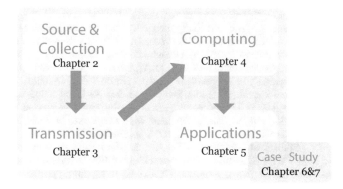

Fig. 1.2 The monograph organization

1.3 Organization of the Monograph

The organization of this monograph follows the life cycle of the mobile big data as shown in Fig. 1.2. The data generation, data sources and data collection are discussed in Chap. 2. The supporting infrastructure of mobile big data for transmissions will be explored in Chap. 3. In Chap. 4, we will discuss the hardware and software platforms for big data processing, which is the critical component to facilitate mobile big data driven applications. The latter, together with related methodologies, are reviewed in Chap. 5. In Chaps. 6 and 7, two case studies [37, 38] are presented based on a real-world network-level mobile dataset, which is employed to study demand forecasting for predictive mobile network management and mobile privacy assessment in terms of user identification across two datasets, respectively.

References

1. X. Cheng, L. Fang, X. Hong, and L. Yang, "Exploiting mobile big data: Sources, features, and applications," *IEEE Network*, vol. 31, no. 1, pp. 72–79, Jan. 2017.
2. Y. Cao, P. Hou, D. Brown, J. Wang, and S. Chen, "Distributed analytics and edge intelligence: Pervasive health monitoring at the era of fog computing," in *Proceedings of the 2015 Workshop on Mobile Big Data*, Hangzhou, China, Jun. 22–25, 2015, pp. 43–48.
3. R. LiKamWa, Y. Liu, N. D. Lane, and L. Zhong, "Moodscope: Building a mood sensor from smartphone usage patterns," in *Proceedings of the 11th Annual International Conference on Mobile Systems, Applications, and Services*, Taipei, Taiwan, Jun. 25–28, 2013, pp. 389–402.
4. Q. Han, S. Liang, and H. Zhang, "Mobile cloud sensing, big data, and 5G networks make an intelligent and smart world," *IEEE Network*, vol. 29, no. 2, pp. 40–45, Mar. 2015.
5. D. Laney, "3D data management: Controlling data volume, velocity and variety," *META Group Research Note*, vol. 6, p. 70, Feb. 2001.
6. X. Dong and D. Srivastava, "Big data integration," in *Proceedings of the 29th IEEE International Conference on Data Engineering (ICDE)*, Brisbane, Australia, Apr. 8–12, 2013, pp. 1245–1248.

7. J. Gantz and D. Reinsel, "Extracting value from chaos," *IDC iView*, no. 1142, pp. 9–10, 2011.
8. M. Ficek and L. Kencl, "Inter-call mobility model: a spatio-temporal refinement of call data records using a Gaussian mixture model," in *Proceedings of IEEE International Conference on Computer Communications (INFOCOM)*, Orlando, FL, Mar. 25–30, 2012, pp. 469–477.
9. A. Ladd, K. Bekris, G. Marceau, A. Rudys, D. Wallach, and L. Kavraki, "Using wireless Ethernet for localization," in *Proceedings of the IEEE/RSJ International Conference on Intelligent Robots and Systems*, Lausanne, Switzerland, Sep. 30–Oct. 4, 2002, pp. 402–408.
10. B. Ferris, D. Fox, and N. D. Lawrence, "WiFi-SLAM using Gaussian process latent variable models," in *Proceedings of the 20th International Joint Conference on Artificial Intelligence (IJCAI)*, vol. 7, Hyderabad, India, Jan. 6–12, 2007, pp. 2480–2485.
11. J. Huang, D. Millman, M. Quigley, D. Stavens, S. Thrun, and A. Aggarwal, "Efficient, generalized indoor WiFi GraphSLAM," in *Proceedings of IEEE International Conference on Robotics and Automation (ICRA)*, Shanghai, China, May 9–13, 2011, pp. 1038–1043.
12. M. Balakrishnan, I. Mohomed, and V. Ramasubramanian, "Where's that phone?: geolocating IP addresses on 3G networks," in *Proceedings of the 9th ACM SIGCOMM conference on Internet Measurement Conference*, Chicago, Illinois, Nov. 4–6, 2009, pp. 294–300.
13. A. Metwally and M. Paduano, "Estimating the number of users behind IP addresses for combating abusive traffic," in *Proceedings of the 17th ACM International Conference on Knowledge Discovery and Data Mining*, San Diego, CA, Aug. 21–24, 2011, pp. 249–257.
14. L. T. Le, T. Eliassi-Rad, F. Provost, and L. Moores, "Hyperlocal: Inferring location of IP addresses in real-time bid requests for mobile Ads," in *Proceedings of the 6th ACM SIGSPATIAL International Workshop on Location-Based Social Networks*, Orlando, FL, Nov. 5, 2013, pp. 24–33.
15. H. Hu and D.-L. Lee, "Semantic location modeling for location navigation in mobile environment," in *Proceeding of IEEE International Conference on Mobile Data Management (MDM)*, Berkeley, CA, Jan. 19–22, 2004, pp. 52–61.
16. B. Shaw, J. Shea, S. Sinha, and A. Hogue, "Learning to rank for spatiotemporal search," in *Proceedings of the 6th ACM International Conference on Web Search and Data Mining*, Rome, Italy, Feb. 4–8, 2013, pp. 717–726.
17. J. Goncalves, S. Hosio, D. Ferreira, and V. Kostakos, "Game of words: Tagging places through crowdsourcing on public displays," in *Proceedings of the 2014 Conference on Designing Interactive Systems*, Vancouver, Canada, Jun. 7–11, 2014, pp. 705–714.
18. N. Stojanovic, L. Stojanovic, Y. Xu, and B. Stajic, "Mobile CEP in real-time big data processing: Challenges and opportunities," in *Proceedings of the 8th ACM International Conference on Distributed Event-Based Systems*, Mumbai, India, May 26–29, 2014, pp. 256–265.
19. M. Mardani, G. Mateos, and G. B. Giannakis, "Subspace learning and imputation for streaming big data matrices and tensors," *IEEE Transactions on Signal Processing*, vol. 63, no. 10, pp. 2663–2677, May 2015.
20. A. Tajer, V. V. Veeravalli, and H. V. Poor, "Outlying sequence detection in large data sets: A data-driven approach," *IEEE Signal Processing Magazine*, vol. 31, no. 5, pp. 44–56, Sep. 2014.
21. D. Z. Yazti and S. Krishnaswamy, "Mobile big data analytics: Research, practice, and opportunities," in *Proceedings of the 15th IEEE International Conference on Mobile Data Management (MDM)*, Brisbane, QLD, Jul. 14–18, 2014, pp. 1–2.
22. L. Sweeney, "K-anonymity: A model for protecting privacy," *International Journal of Uncertainty, Fuzziness and Knowledge-Based Systems*, vol. 10, no. 5, pp. 557–570, Oct. 2002.
23. H. Zang and J. Bolot, "Anonymization of location data does not work: A large-scale measurement study," in *Proceedings of the 17th Annual International Conference on Mobile Computing and Networking*, Las Vegas, Nevada, USA, Sep. 19–23, 2011, pp. 145–156.
24. Y.-A. de Montjoye, C. A. Hidalgo, M. Verleysen, and V. D. Blondel, "Unique in the crowd: The privacy bounds of human mobility," *Scientific Reports*, vol. 3, Mar. 2013.
25. M. Gramaglia, M. Fiore, A. Tarable, and A. Banchs, "Preserving mobile subscriber privacy in open datasets of spatiotemporal trajectories," in *Proceedings of IEEE International Conference on Computer Communications (INFOCOM)*, Atlanta, GA, USA, May 1–4, 2017, pp. 1–9.

26. F. M. Naini, J. Unnikrishnan, P. Thiran, and M. Vetterli, "Where you are is who you are: User identification by matching statistics," *IEEE Transactions on Information Forensics and Security*, vol. 11, no. 2, pp. 358–372, Feb. 2016.
27. C. Riederer, Y. Kim, A. Chaintreau, N. Korula, and S. Lattanzi, "Linking users across domains with location data: Theory and validation," in *Proceedings of the 25th International Conference on World Wide Web*, Montreal, Quebec, Canada, Apr. 11–15, 2016, pp. 707–719.
28. A. Narayanan and V. Shmatikov, "Robust de-anonymization of large sparse datasets," in *Proceedings of IEEE Symposium on Security and Privacy*, Oakland, CA, May 18–22, 2008, pp. 111–125.
29. Y. De Mulder, G. Danezis, L. Batina, and B. Preneel, "Identification via location-profiling in GSM networks," in *Proceedings of the 7th ACM Workshop on Privacy in the Electronic Society*, Alexandria, Virginia, USA, 2008, pp. 23–32.
30. A. Cavoukian, "Introduction to privacy by design," 2016. [Online]. Available: https://www.ipc.on.ca/english/Privacy/Introduction-to-PbD/
31. J. K. Laurila, D. Gatica-Perez, I. Aad, J. Blom, O. Bornet, T.-M.-T. Do, O. Dousse, J. Eberle, and M. Miettinen, "The mobile data challenge: Big data for mobile computing research," in *Proceedings of Nokia Workshop on Mobile Data Challenge in conjunction with International Conference on Pervasive Computing*, Newcastle, UK, Jun. 18–20, 2012.
32. N. Kiukkonen, J. Blom, O. Dousse, D. Gatica-Perez, and J. Laurila, "Towards rich mobile phone datasets: Lausanne data collection campaign," in *Proceedings of ACM International Conference on Pervasive Services*, Berlin, German, Jul. 13–16, 2010, pp. 1–7.
33. I. Aad and V. Niemi, "NRC data collection and the privacy by design principles," *Phone Sense*, pp. 41–45, Nov. 2010.
34. M. Musolesi, "Big mobile data mining: Good or evil?" *IEEE Internet Computing*, vol. 18, no. 1, pp. 78–81, Jan. 2014.
35. M. L. Damiani, E. Bertino, and C. Silvestri, "The PROBE framework for the personalized cloaking of private locations," *ACM Transactions on Data Privacy*, vol. 3, no. 2, pp. 123–148, Aug. 2010.
36. M. Damiani, C. Silvestri, and E. Bertino, "Fine-grained cloaking of sensitive positions in location-sharing applications," *IEEE Pervasive Computing*, vol. 10, no. 4, pp. 64–72, Apr. 2011.
37. L. Fang, X. Cheng, H. Wang, and L. Yang "Mobile Demand Forecasting via Deep Graph-Sequence Spatiotemporal Modeling in Cellular Networks," *IEEE Internet of Things Journal*, vol. 99, no. 99, 2018.
38. L. Fang, H. Wang, X. Cheng, and L. Yang "Mobile Privacy: User Identification via Ensemble Matching on Spatiotemporal Features," *IEEE Transactions on Information Forensics and Security*, vol. 99, no. 99, 2018.

Chapter 2
Source and Collection

2.1 Overview of Data Sources

Mobile data can be collected from various sources in the mobile network. These data are usually divided into two categories [1]. One category consists of the *app-level* data directly collected by mobile App vendors from mobile phone sensors. As sensor technologies are ubiquitously equipped in smart phones (e.g., GPS, accelerometer, magnetic field sensor, gyroscope, etc.), the phone usually acts as a sensor hub with enriched connectivity for data collection and transmission. The other data category is the *network-level* one traditionally collected by content service providers and mobile operators, which is a vast amount of various mobile service contents, as well as spatiotemporal mobile broadband data about their systems and customers. This type of data records the system status, the service requests, as well as user information (e.g., user ID, location, device type, time stamps, type of service, etc.).

In terms of the sources of data collection, the app-level data mainly come from the mobile terminals, whereas the network-level data are usually from the over the top (OTT) servers and the network operators. The raw data collected from these sources is summarized in Fig. 2.1. Embedded in these raw data is a large amount of valuable information about the users, including user characteristics, habits, preferences, and even motivations and purposes. Harvesting from these raw data, one can construct more useful information such as context, behavior, relationship, etc. Based on these, additional and more implicit information can be further extracted via data mining. Examples include: basic user characteristics (age, gender, race), occupation, group, habit, interest, political opinion, etc. These could then be used in followup data analytics to restore the original context of the related mobile terminal utilization.

	Data	Parameter
app-level data	Device	Device Type, Device Usage, etc.
	Profile	MSISDN, IMEI, IMSI, User preference, Calendar, Appointment, etc.
	Sensor	Sensing Data, e.g., GPS, Gyroscope, Accelerometer, etc.
	App	Terminal Application Type, Application Usage, etc.
	Service	Service Information, e.g., Bundle Type, Service Charge, etc.
	Log	Terminal Device Log, Server System Log, etc.
network-level data	Time	Connection Starting Time, Session Starting Time, etc.
	Location	Terminal Location, BS Location, Router Location, Cell Location, etc.
	Address	Client IP, Server IP, Client Tunnel IP, Server Tunnel IP, etc.
	URL	Uniform Resource Location, Link Information, Link Content, etc.
	Flow	Uplink traffic, Downlink Traffic, Packet Number, etc.
	Record	Conversation Log, e.g., Conversation Duration, Conversation Time, Conversation Frequency, etc.

Fig. 2.1 Basic data and parameters

Data collection is the process in which data containing user characteristics, preferences, or activities is obtained. The manner in which the data collection is implemented can be classified into implicit and explicit approaches. In the explicit approach, users are prompted to manually provide various information [2–5]. While being simple and straightforward, this requires each user to be not only clear about what relevant information he/she is disclosed, but also willing to take time and effort to participate. However, this is usually hard to achieve, as users could be discouraged by such inquiries. On the contrary, the implicit approach does not require manual user intervention and is accomplished without interfering with normal user activities. The implicit approach also facilitates more frequent information updates since explicit user responses are not required in such updates. For these reasons, the implicit approach is more prevalent. Nevertheless, implicitly collected data usually contains quite a lot of redundancy and irrelevant information, which could complicate the followup processing of the data. In the following subsections, we will present the data in terms of app level and network level.

2.1.1 The App-Level Data

Data collected from mobile devices may be from either the software side or the hardware side. The hardware-side data includes the device usage information, sensor information, etc. The software-side data includes the application information, the user profile associated with the devices, and the system logs [6]. There have been quite a few projects focusing on the collection of data from the mobile terminals. Reality mining carried out by the MIT Human Dynamics Lab over 9 months in 2004 was among the earliest efforts, where 75 faculty and students with the MIT Media Lab and 25 students at the MIT Sloan business school, participated using 100 Nokia 6600 smart phones [7]. In this experiment, call logs, bluetooth devices in proximity, cell tower IDs, phone status (charging or idle), and popular application usage data have been collected. In the more recent Mobile Data Challenge (MDC) by Nokia, 200 volunteers participated using Nokia N95 in the Lake Geneva region from October 2009 to March 2011 [8]. Data collected include calls, short messages, photos, videos, application events, calendar entries, location points, historically connected cell towers, accelerometer samples, Bluetooth observations, historically connected Bluetooth devices, WLAN observations, historically connected WLAN access points and audio samples. Since March 2011, the Device Analyzer experiment at a much larger scale involving 12,500 Android devices was carried out by the Computer Laboratory at the University of Cambridge [9, 10]. The records of covered countries, phone types, OS versions, device settings, installed applications, system properties, bluetooth devices, WiFi networks, disk storage status, energy and charging status, telephony, data usage, CPU and memory status, alarms, media and contacts, as well as sensors have been collected and analyzed. These campaigns have been summarized in Fig. 2.2.

2.1.2 The Network-Level Data

These data are typically collected either at the OTT servers or at the network operator servers. The raw information at the OTT servers consists of a vast amount of texts, user profiles, system logs, audio and visual contents etc. Most of OTT service providers directly interact with end users, rendering network operators pure "pipes," and thus keeping them away from the invaluable data flow.

On the other hand, the radio access network data mainly come from the interactions between mobile terminals and base stations, which involve cell search, synchronization, link establishment, uplink and downlink data transfer, handover, and system information broadcast. These lead to the exchange of a variety of data involving multiple network layers, such as network and device identity, power/carrier/antenna indices, payload and transmission mode,

Project	Time	Organization	Data Collected
Reality Mining http://realitycommons.media.mit.edu/	2004	MIT Human Dynamic Lab	call logs, Bluetooth devices in proximity, cell tow IDs, phone status, popular application usage data
Mobile Data Challenge (MDC) https://www.idiap.ch/dataset/mdc	2009-2011	Nokia	calls, SMS, photos, videos, application events, calendar entries, location points, unique cell towers, accelerometer samples, etc.
Device Analyzer Experiment https://deviceanalyzer.cl.cam.ac.uk/	2011 - ~	Computer Laboratory at the University of Cambridge	covered countries, phone types, OS versions, device settings, installed applications, system properties, Bluetooth devices, WiFi networks, disk storage, energy and charging, telephony, data usage, CPU and memory, alarms, media and contacts

Fig. 2.2 Summary of mobile data collection projects

timing information, and location. Details of data collection by network operators will be discussed in next section.

Compared with the data from the content service providers and mobile terminal devices, the server data items unique to network operators include: location, address, time, record, flow, URL etc. Among these, "location" contains the locations of the base stations (location area code, LAC), the cells (service area code, SAC) and the routers (routing area code, RAC), from which each individual user's physical position could be uniquely determined, without the assistance of the mobile terminal GPS. "Address" contains the IP addresses of the clients, the servers, and the tunnels, etc. "Time" contains the starting time stamps of user's connections and sessions. Also uniquely accessible by the network operators are the user mobile number (MSISDN) and user device identity (IMEI), from which each individual user's specific device can be determined. These data, being privacy sensitive, are not typically accessible by other sources of data collection, unless voluntarily provided by the users. The latter case, however, could potentially compromise the reliability of collected data depending on the user's true willingness to disclose such data.

2.2 Data Collection in Mobile Networks

In this section, the architecture of mobile networks and key network components as well as the mobility management mechanism are first reviewed, based on which the revealed user network behaviors could be better understood. Then, the data collection and data categorization based on the heterogeneous data collection points in cellular networks are described and discussed in detail.

2.2.1 Network Architecture Overview

The mobile (cellular) network emerged in the 90s of last century and has become one of the most successful technologies. The original cellular network is aimed to provide voice service wirelessly by distributing multiple base stations within a covered area, each of which is covering a small region exclusively (abstracted as a hexagon in Fig. 2.3). The data traffic capability was added to cellular networks from the second generation of cellular networks and flourished in the fourth generation, the long-term evolution (LTE). Although cellular networks have significantly evolved since its first generation, its two main components remain the same, namely the radio access networks (RAN) and the core networks (CN). In a cellular network, the RAN is responsible for processing wireless signals (baseband and passband) from user equipments (UEs), while the CN is aimed to reliably direct the outgoing and incoming traffic flow to their respective destinations.

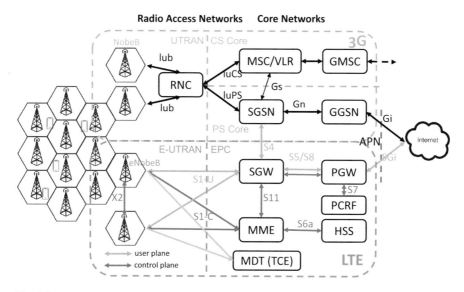

Fig. 2.3 Cellular network architecture overview (3G and LTE)

In general, two trends of the cellular network architecture evolution can be observed, namely the packetization and the user-control plane separation. Such cellular network evolution trends largely improve the network delay performance and capacity, which could also facilitate accessible network-level data collection. In 3G networks, data traffic service in the core network is accomplished by the packet switched core (PS Core), while the traditional voice and texting services are fulfilled by the legacy circuit switched core (CS Core) inherited from the 2G cellular networks. However, all services in the LTE cellular networks are fulfilled via the evolved packet core (EPC), which could simplify the system architecture and enhance system efficiency. Data collection in mobile networks can also benefit from packetization, as the packetized networks could provide more bandwidth for big data collection and transmission.

The other trend of cellular network evolution is the user-control plane separation. In general, the user plane in a network refers to the network that carries data traffic, while the control plane is the network for controlling signal transmissions. In LTE networks, the user-control plane on the interfaces between E-UTRAN and EPC is first separated (interfaces S1-C and S1-U in Fig. 2.3), and then the interface between the serving gateway (SGW) and the packet data network gateway (PGW) (interface S5 (internal)/S8 (roaming) in Fig. 2.3) in 3GPP LTE Standard Release 14. The user-control plane separation could generally reduce the network delay via a centralized control function and support the increase of data traffic by adding user plane nodes without changing the network controlling components. At the same time, the user-control plane separation can also facilitate collection of user data related to the distinct network behaviors.

As LTE consists of the main stream of mobile networks nowadays, the mobile network architecture will be illustrated from the perspective of LTE, and the counterparts of the network functionalities in 3G networks will be briefly introduced. In Fig. 2.3, the network architectures of both 3G and LTE (4G) cellular networks are plotted. The double-arrow lines in the figure refer to the logical network connection, beneath which physical transport networks, typically IP networks, are employed to fulfill the network logical connections. In addition, it is worth noting that a logical connection may not necessarily imply a direct physical connection. For example, the interface among nearby eNodeBs, X2, is not necessarily implemented as direct physical connections, but can be achieved by routing through the core network.

2.2.2 Key Network Components

The architecture of LTE is outlined in Fig. 2.3, and the main components therein are introduced as follows:

Evolved Node B The evolved node B (eNodeB or eNB) represents base stations covering user equipments (UEs) in a certain area, via which an UE can only

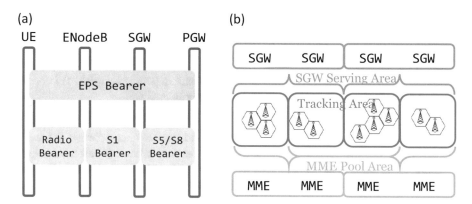

Fig. 2.4 Bearer and various networks area definition in the LTE. (**a**) User-plane bearers, (**b**) Network area

communicate with and reach the remote destination. The eNobeB has two main functions. The first function is to process the uplink and downlink radio signals via analogue and digital signal processing, while the other one is to fulfill low-level controls via signaling messages (e.g., handover). In fact, the low-level control functions of eNodeB in LTE are inherited from the radio network controller (RNC) in 3G networks as shown in Fig. 2.3, which could reduce the delay due to the reduction of control message exchanges between RNC and base stations. Each eNodeB is connected to EPC via interface S1 and to nearby eNodeBs via interface X2.

Tracking Area (TA) To facilitate effective system and user management, especially for mobility management, the entire covered area is partitioned into multiple tracking areas (TA), each of which is exclusively comprised of several base stations (eNodeBs) spatially adjacent to each other. In fact, the TA serves as a basic geographic unit for the service coverage area of network components as shown in Fig. 2.4b. In addiction, the TA is also the basic location unit for user mobility management in LTE networks, when users are in the idle state.

Mobility Management Entity (MME) Mobility management entity (MME) is the critical controlling component in LTE networks, which is the main signaling node in the EPC control plane. Some control functionalities of the MME are inherited from the RNC in 3G networks. In the initial UE attaching phase (UE switch on), the MME will first authenticate and authorize the UE by cooperating with the home subscriber server (HSS) and then assign a proper serving gateway (SGW) to serve the UE. The load of SGWs is also balanced by the MME by directing UE from a heavy-loaded SGW to the light-loaded one. Also, the MME keeps tracking the location of each assigned UE at the granularity of TAs in their idle state (details provided in next subsection). Based on the location information of UEs, the MME is also responsible for waking up idle UEs, termed as paging in the context of mobile networks, when an incoming flow for the UE arrives at the associated MME. In fact, the MME is the component in the LTE network that could monitor user spatiotemporal behaviors, regardless of the UE status (active or idle). This could potentially provide tremendous value to the data collected here.

Serving Gateway (SGW) The serving gateway acts as a high-level router, forwarding the data (user) traffic between eNodeBs and packet data network gateways (PGWs). A network typically contains many serving gateways, each of which handling UEs in a geographical area in terms of TAs. The latter is termed as the SGW serving area, which is not necessarily exactly the same as MME pool area (as shown in Fig.2.4b). The SGW is also responsible for inter-eNodeB handovers in the user plane to seamlessly direct data traffic from the outdated eNodeB to the updated one. The downlink traffic for an idle UE is also buffered at the SGW, before the idle UE is woken up via the paging procedure scheduled by the MME.

Packet Data Network Gateway (PGW) The packet data network gateway (PGW) is the point of connection between the PC and external IP networks via interface SGi. Each packet data network (PDN) can be pinpointed by an identifier termed as the access point name (APN). Each UE will be assigned a default PGW in its switch-on initialization. The latter could be attached to other PDNs for private accesses. Typically, the HSS holds a PDN list that a UE can connect to. In fact, PGWs are also responsible for packet filtering, charging support, QoS rule and policy enforcement, which is fulfilled by the policy control enforcement function (PCEF). Generally, the PCEF resides in the PGW and is connected to the policy and charging rule function (PCRF) via interface S7, which is responsible for policy control decision-making and the flow-based charging functionality. In fact, PCRF could be viewed as a data aggregation combining device, network, location and billing information of subscribers. Clearly, PCRF is a typical data collection point in cellular networks.

Bearers In LTE, the logical connection between two nodes in the EPC is termed as the bearer (session). It could be viewed as a bidirectional tunnel. The bearer is designed to address the special issues in LTE networks, namely mobility and quality of service control. In fact, two types of bearers are defined in LTE networks, namely control-plane (signaling) bearers and user-plane (data traffic) bearers. In Fig. 2.4a, the user-plane bearer from UE to PGW is illustrated. In fact, a default evolved packet system (EPS) bearer will be assigned to UEs in their switch-on initialization, which provides a tunnel for UEs to communicate with external networks. The EPS bearer is comprised of three low-level bearers, each of which corresponding to a specific interface. The resultant bearers include the radio bearer, the S1 bearer and the S5/S8 bearer. The activation of these bearers relies on the network behaviors of UEs, which will be discussed in following subsection.

2.2.3   Mobility Management and User Network Behaviors

The network behavior of users does not remain unaltered when it registers and is attached to a LTE network. The state of network users is defined by the network behavior diagram shown in Fig. 2.5. Such user management mechanism is aimed to address the issues of limited UE battery life and signaling traffic overload in LTE networks. Once a UE is attached to a LTE network in its switch-on initialization, the UE enters the EPS mobility management (EMM) REGISTERED state from the DEREGISTERED state. At the same time, the UE will be assigned a serving MME, a serving SGW and a default EPS bearer, based on the UE location and the load status of the available MMEs and SGWs. In this phase, the UE enters the EMM/RCC CONNECTED state, indicating that the UE has the full connectivity to the external world. The radio resource control (RCC) state is the one viewed from the perspective of RANs, while the EMM one is viewed from the EPC. Generally, these two states are equivalent. In the EMM/RCC CONNECTED state, the MME has the UE's location information at the granularity of eNobeB. That is, the MME knows the exact eNodeB the UE is attached to as long as the UE is in the EMM/RCC CONNECTED state. It is also worth noting that UEs in the EMM/RCC CONNECTED state will trigger a handover (HO) event when it arrives a new cell, so that the ongoing service could be seamlessly transferred from the outdated eNodeB to the new one.

When the UE is registered but does not consume any radio resources for any services, the S1 release procedure will be scheduled to shift the UE into the EMM/RCC IDLE state. The S1 release procedure is initialized by the UE-attached eNodeB to release the assigned radio bearer and S1 bearer resources. However, the S5/S8 bearer will be retained to accept the UE's downlink data traffic from the external networks. In the EMM/RCC IDLE state, the UE could freely move around with limited signaling message exchanges with eNodeBs and EPC. Also, the MME only has the location knowledge of the UE at the granularity of tracking areas. To facilitate mobility management in LTE, tracking area updates will be triggered by two events to maintain the MME's knowledge of the registered UEs' status and

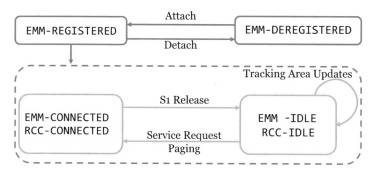

Fig. 2.5 User network behaviors

location. The first event is that the UE enters a new tracking area that is not in the UE's recent tracking area list. The second event is the expiration of periodic tracking area update timer, whose time duration is typically 54 min but can be customized by network operators. During a tracking area update procedure, the UE will temporally enter the EMM/RCC CONNECTED state and then finally return to the EMM/RCC IDLE state.

The transition from the EMM/RCC IDLE state to the EMM/RCC CONNECTED state of UEs is triggered by two events. First, the incoming flow to the UE arrives at the serving SGW via interface S5/S8. The paging procedure is triggered by the SGW and scheduled by the MME to search and wake the UE up within the latest tracking area updated by the UE. During the paging procedure, the radio and S1 bearers will be re-assigned to the UE so that the connection between the UE and the external networks could be established. Thus, the UE's state changes from IDLE to CONNECTED. Secondly, the UE will initialize a service request procedure when it has a communication demand. The service request procedure will sequentially re-establish the radio bearer and S1 bearer at the eNodeB and the serving SGW, respectively. As a result, the UE's state is changed to CONNECTED so that the UE could communicate with external networks.

2.2.4 Data Collection and Categorization

Based on the previous description of network architecture and user network behaviors, the characteristics of data collected at different spots of mobile networks will be discussed here. Generally, four types of dataset could be categorized for the network-level data collected in mobile networks, namely the call detail records (CDRs) data, the user-plane traffic (UPT) data, the control-plane traffic (CPT) data and the radio measurement reports (RMR) data, as summarized in Fig. 2.6.

Call Detail Records (CDR) The CDR data is the most popular dataset studied in the literature [11, 12]. Originally collected for service charging purposes by network operators, the CDR data typically record users' voice and texting activities. Its data fields include the user identifier, when (time stamp) and where (at the granularity of base stations) the event occurs, the duration that the event lasts for voice service. The CDR data may also include the data traffic volume consumed by each UE. The reason behind the high popularity of CRD data is the high accessibility of such data, as the CDR data typically resides at a single server and is well structured. However, the CDR data can only provide the user information for users in the CONNECTED state. Users in the IDLE state do not generate any input to the CDR data. In addition, users' data traffic behaviors may be difficult to be thoroughly monitored based on the CDR data alone.

User-Plane Traffic (UPT) Data The UPT data refers to the IP session data collected at the PGW (LTE) or GGSN(3G) of cellular networks. The UPT data is generated by inspecting messages tunneled in GRPS tunneling protocol (GTP-U). It encapsulates

	CDR	UPT	CPT	RMR
Collection Point	Charging Station	PGW (LTE) / GGSN (3G)	MME (LTE) / MSC (3G)	Base Stations / MDT (LTE)
Data Fields	User ID Service Type Time Start Location Duration (voice) Volume (traffic)	User ID Session Start Time Session End Time Session Start Location Uplink Volume Downlink Volume Upper Layer Protocols	User ID Event Type Time Location	User ID Time Location WCQI Serving RSRP Serving RSRQ RB Load
Location Granularity	Cells	Cells	Cells Tracking Area	GPS Coordinates
Sampling Rate	Per Service	Per Data Traffic Service	Per Service Per Periodic TAU (default: 54 min)	Determined by Network Operator
Participated Users	All	All	All	Limited (Per User's Permission)

Fig. 2.6 Summary of data collections in cellular networks

the IP traffic between UEs and the external networks. The UPT data fields generally include the IP session start and end time, device/user pseudo identifier, type of service, and uplink/downlink traffic volume. Occasionally, the UPT might also include the location of UE at the session start time. However, the user location information in UPT data is sparse and less accurate, as the duration of session will allow users to move to new cells without any records updated in the UPT data.

Control-Plane Traffic (CPT) Data   The CPT data refers to the data collected at controlling components in mobile networks. Signaling data is a typical data type collected at the mobile switching center (MSC) in the CS core of 3G networks or at the MME in the EPC of LTE networks. In LTE, the data collected at the MME could have a higher observability on UE mobility behaviors, compared with the location information of CDR and UPT data. Based on the network mobility management mechanism in LTE networks discussed previously, the MME has the knowledge of the UE location at the granularity of cells when the UE is in the CONNECTED state. Even when UEs are in the IDLE state, the MME still knows their location at the granularity of tracking area via the tracking area updating mechanism of mobility management. In fact, tracking area updates provide the location information in terms of cells at which UEs report their locations. Furthermore, the periodic tracking area update frequency could be significantly increased from a 54-min update interval to a 14-min one [13], providing more detailed and more accurate observations on UE mobility behaviors. The data collected at the MSC of 3G networks also has the records of UEs' voice and texting service activities. The data fields of CPT data typically include the user identifier, event type, cell ID, and time stamp, etc.

Radio Measurement Reports (RMR) The RMR refers to the data based on radio measurement reports generated at UEs. It is originally aimed to facilitate radio network operation and radio network performance assessments. The RMR is generally difficult to collect, due to the distributed nature of base stations and UEs. In addition, the limited storage and computation capabilities of base stations also limit the availability of the RMR data. A typical example of RMR data is the measurement reports collected from the minimization of drive tests (MDT) server. The MDT functionality [14] is originally designed in LTE standards to collect radio measurement reports directly from UEs to minimize the drive testing of network operators for radio network performance assessments. The data fields of MDT data typically include the user ID, wideband channel quality indication (WCQI), serving reference signal received power (RSRP) and quality (RSRQ), as well as resource block (RB) load [15]. Occasionally, the user throughput is also included in the MDT data. The location information of UEs is provided by their GPS receivers at the granularity of meters, which results in much more precise location observations at the intra-cell level, in comparison with other data. However, such data collection requires the permission of UEs and the investment of infrastructure for data collection and transmission, both of which will limit the availability of RMR data.

References

1. D. Z. Yazti and S. Krishnaswamy, "Mobile big data analytics: Research, practice, and opportunities," in *Proceedings of the 15th IEEE International Conference on Mobile Data Management (MDM)*, Brisbane, QLD, Jul. 14–18, 2014, pp. 1–2.
2. Y. Park and E. Lee, "A new generation method of a user profile for information filtering on the internet," in *Proceedings of the 12th International Conference on Information Networking*, Tokyo, Japan, Jan. 21–23, 1998, pp. 261–264.
3. R. J. Mooney and L. Roy, "Content-based book recommending using learning for text categorization," in *Proceedings of the 15th ACM Conference on Digital Libraries*, San Antonio, TX, 2000, pp. 195–204.
4. M. Pazzani, J. Muramatsu, and D. Billsus, "Syskill & webert: Identifying interesting web sites," in *Proceedings of the Thirteenth National Conference on Artificial Intelligence*, Portland, OR, Aug. 4–8, 1996, pp. 54–61.
5. W. Kim, L. Kerschberg, and A. Scime, "Learning for automatic personalization in a semantic taxonomy-based meta-search agent," *Electronic Commerce Research and Applications*, vol. 1, no. 2, pp. 150–173, Summer 2002.
6. C. C. Tossell, P. Kortum, C. W. Shepard, A. Rahmati, and L. Zhong, "Getting real: a naturalistic methodology for using smartphones to collect mediated communications," *Human-Computer Interaction*, vol. 2012, no. 10, pp. 1–10, Apr. 2012.
7. E. Nathan and A. Pentland, "Reality mining: sensing complex social systems," *Personal and Ubiquitous Computing*, vol. 10, no. 4, pp. 255–268, Mar. 2006.
8. J. K. Laurila, D. Gatica-Perez, I. Aad, J. Blom, O. Bornet, T.-M.-T. Do, O. Dousse, J. Eberle, and M. Miettinen, "The mobile data challenge: Big data for mobile computing research," in

Proceedings of Nokia Workshop on Mobile Data Challenge in conjunction with International Conference on Pervasive Computing, Newcastle, UK, Jun. 18–20, 2012.

9. D. Wagner, A. Rice, and A. Beresford, "Device analyzer: Understanding smartphone usage," in *Proceedings of the 10th International Conference on Mobile and Ubiquitous Systems: Computing, Networking and Services*, Tokyo, Japan, Dec. 2–4, 2013, pp. 195–208.

10. D. T. Wagner, A. Rice, and A. R. Beresford, "Device analyzer: Large-scale mobile data collection," *ACM SIGMETRICS Performance Evaluation Review*, vol. 41, no. 4, pp. 53–56, Mar. 2014.

11. S. Han, C. L. I, G. Li, S. Wang, and Q. Sun, "Big data enabled mobile network design for 5g and beyond," *IEEE Communications Magazine*, vol. 55, no. 9, pp. 150–157, Jul. 2017.

12. D. Naboulsi, M. Fiore, S. Ribot, and R. Stanica, "Large-scale mobile traffic analysis: A survey," *IEEE Communications Surveys Tutorials*, vol. 18, no. 1, pp. 124–161, Firstquarter 2016.

13. Q. Lv, Y. Qiao, N. Ansari, J. Liu, and J. Yang, "Big data driven hidden Markov model based individual mobility prediction at points of interest," *IEEE Transactions on Vehicular Technology*, vol. 66, no. 6, pp. 5204–5216, Jun. 2017.

14. J. Johansson, W. A. Hapsari, S. Kelley, and G. Bodog, "Minimization of drive tests in 3GPP release 11," *IEEE Communications Magazine*, vol. 50, no. 11, pp. 36–43, Nov. 2012.

15. F. Chernogorov and J. Puttonen, "User satisfaction classification for minimization of drive tests QoS verification," in *Proceedings of the 24th IEEE Annual International Symposium on Personal, Indoor, and Mobile Radio Communications (PIMRC)*, London, UK, Sep. 8–11, 2013.

Chapter 3
Transmission

The collection, transmission, and computing of mobile big data require the support of communication, networking and computing infrastructure. Due to the special characteristics of mobile big data, the communications and networking infrastructure urges a revolutionary overhaul. For example, the (near) real-time response demanded by some mobile big data driven applications is hardly satisfied by the existing infrastructure. In this section, we survey the potential technologies on communications, transmissions and computing in the context of mobile big data.

Research challenges on the infrastructure supporting mobile big data are always entangled with the tradeoff between centralization and distribution of resource management and system design. Specifically, centralization brings efficiency and convenience to the system management and coordination, but falls short in terms of scalability. On the other hand, distribution usually leads to improved scalability, but lacks the easiness on global system management and coordination. Hence, the issue of how to design the system to support mobile big data collection, processing and sharing, considering the tradeoff between centralization and distribution, is always of great interest, which will be discussed in the following sections.

3.1 Computing Infrastructure

3.1.1 Mobile Cloud Computing

The concept of centralized mobile cloud computing (MCC) [1, 2] (Fig. 3.1) is proposed to solve the problem of mobile big data processing, by integrating mobile sensing and cloud computing. The intensive computing workload and high-volume data storage demand of mobile big data processing are loaded to the cloud via certain access and backhaul networks.

© Springer International Publishing AG, part of Springer Nature 2018
X. Cheng et al., *Mobile Big Data*, Wireless Networks,
https://doi.org/10.1007/978-3-319-96116-3_3

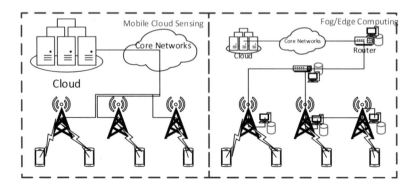

Fig. 3.1 Computing paradigms to support mobile big data

With the idea of MCC, the bottleneck of mobile big data processing is shifted to the communication between the mobile devices and the cloud. The involved access and backhaul connections should be able to handle massive data transmissions due to the tremendous volume of mobile big data, as well as massive simultaneous device connection requests.

There are some major challenges to apply MCC for mobile big data processing. First, the current radio access networks may not be able to meet the intensive future needs of mobile big data transmissions. In addition, the MCC needs to adapt to the randomly varying communication quality, low security and high probabilities of signal interception [3]. Secondly, the latency due to access and backhaul networks is a vital challenge [4] for mobile cloud computing, especially when interactions between mobile terminals and the cloud are required in real time. In addition, the degrading communication quality will be intensified by the high latency of the backhaul networks, and such latency is difficult to control in traditional networks, as routers and switches in the traditional computer networks are locally operated and controlled. Hence, how to reduce the high transmission latency in the context of mobile big data poses a great challenge, and recent studies on this issue can be found in [5–7].

3.1.2 Fog/Edge Computing

In order to reduce the network delay coming from the backhaul network, the concept of fog computing [8] (as shown in Fig. 3.1) is proposed to bring the computing and storage capability closer to the mobile devices, near the edge of the network. In other words, devices located at the edge of the Internet, such as routers, switches, base stations, access points, etc., will be equipped with computing and storage resources. In fact, fog computing extends cloud computing schemes from the core of the network to the edge of the network.

In the context of mobile big data, the fog computing paradigm can deal with data acquisition, aggregation and preprocessing, and even data mining, without suffering from the high latency as in mobile cloud computing. However, the computing and storage resources of a single network device at the edge of the network may not have sufficient capability to handle the mobile big data tasks, such that cooperation among edge devices with limited individual computing capability is of great interest. The concept of cloudlet is to form a cloud-like computing paradigm based on multiple edge devices with computing resources in physical proximity, in order to both reduce the latency and provide powerful computing resources via mobile device cooperation [4, 9]. Accordingly, efficient heterogeneous computing resource management in a hierarchical network poses research challenges and provides great research opportunities. In particular, the interaction and coordination control among the edge devices leads to many intriguing research problems.

Although the paradigm of fog computing can reduce the latency to the core of the Internet, the bandwidth and connectivity limitation in the current structure of wireless access networks (especially in the widely used cellular networks) is still present.

3.2 Communication and Networking Infrastructure

In the context of mobile big data, network performance is a key factor that connects mobile terminals and the cloud computing platform. With the development of SDN, network latency may be improved with specific network applications deployed on the centralized control plane. However, there are still challenges in the context of big data applications [10, 11]. For example, the (mobile) big data applications (computing and processing) postulate more rapid and frequent flow table updates, in order to fulfill the needs of bulk data transfer, data aggregation/partition, and so on, in the context of distributed big data computing and storage. This leads to various design and implementation issues in SDN.

3.2.1 Software Defined Networking (SDN)

The difficulty of reducing the latency of the core network largely comes from the distributed nature of the computer network. In fact, network functionalities could be divided into three hierarchical planes: data, control and management [12]. At each network device, the data plane forwards the data packets and the control plane implements the protocols in order to populate the forwarding table for the data plane. The management plane is to monitor and configure the control plane.

In recent years, the idea of software defined networking (SDN) is proposed to cope with the control issue of computer networks, by centralizing the control plane of individual network devices to an external entity (Fig. 3.2a). In other words, the

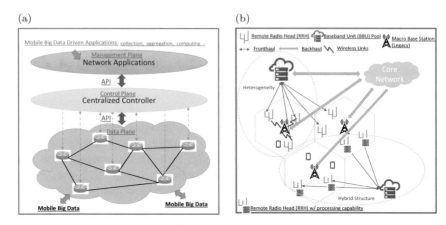

Fig. 3.2 Communications and networking paradigms to support mobile big data. (**a**) Software defined networking (SDN), (**b**) Cloud radio access networks (C-RAN)

data plane is decoupled from the control plane and remotely controlled [12]. With SDN, the forward decisions are based on network flows (defined as a sequence of packets between a source and a destination) rather than the destination of packets. Atop the centralization of the control plane, network applications and services in the management plane, such as routing, firewall, load balancing, status monitoring and so on, are implemented based on programmable interfaces provided by the centralized SDN controller.

3.2.2 Cloud Radio Access Networks (C-RAN)

The unprecedented volume of mobile big data traffic will bring great challenges to current radio access networks (RANs), namely cellular networks in our context, which are generally used in mobile data collection and transmission. The current RAN bandwidth and capacity are not able to fulfill the demand of mobile big data applications. Therefore, the paradigm of RAN needs to be revolutionized.

In the traditional RAN, base stations (BSs) with limited number of antennas can only serve a fixed coverage, which leads to the underutilization of network resources over both space and time. In the evolution of RAN, small cells are preferred to increase the spatial spectrum reuse. However, the interference management and coordination in the hierarchical cell structure post great challenges. In addition, the computing resource in the traditional BSs may not be able to fulfill the demands of dynamic resource management.

The concept of C-RAN [13, 14] is proposed to centralize the computation-intensive functions (baseband processing and resource management) into the back-end cloud connected to BSs via high-capacity connections, which can be wired (like optical fiber) or wireless. Meanwhile, the only function that remains in BSs is the RF-level wireless accessing and possibly some simple symbol processing. Therefore, the radio access networks are essentially divided into two parts, remote radio head (RRH) for RF accessing and baseband unit (BBU) pool for processing, as shown in Fig. 3.2b. The transition from distribution to centralization for baseband processing brings great advantages [2] on load balancing, interference management, multi-cell coordination, etc. In addition, the network device parameters could be reconfigured online, which will provide great flexibility and full utilization of the network resources for high-capacity and reliable radio access services.

With the decoupling between wireless access functions and computing functions, multi-cell joint dynamic resource allocation could be facilitated spatially and temporally in C-RAN, according to the learned user mobility patterns. Furthermore, optimal collaborative radio processing could also be enabled in the centralized computing paradigm of C-RAN, as mobile users could usually be served by multiple small cells. Nevertheless, the design of dynamic resource allocation and collaboration of radio processing to support the real-time high-data-rate applications of mobile big data are facing many open problems.

On the other hand, the computing cloud is able to learn and predict the behaviors of users with the availability of joint spatial and temporal mobile data from the users. The learned knowledge will in turn provide guidance to adjust network structures and reconfigure device parameters, such that the network performance and quality of service can be optimized under the architecture of C-RAN. However, it is challenging to identify and extract useful features from massive mobile big data, as well as to discover the underlying relationship linking mobile user behaviors and network performance.

With the learned knowledge on user behavior, one could cache popular contents in the BSs of macro cells, small cells or even some user devices, which could potentially improve the quality of experience by reducing the content downloading delay, as the content cached at the edge of the network is closer to users. In the literature, caching can be applied not only at the application layer, but also at the network layer [15] or even at the data link layer [16]. However, determination of what to cache is challenging in cache-assisted communication and networking. Generally, the Zipf distribution [17, 18] is assumed to characterize the popularity of contents in most existing results. Although it is well studied that the content popularity follows the Zipf distribution as a whole, it is not accurate to assume that the popularity of contents still follows the Zipf distribution locally in a small region. Therefore, the content popularity as well as the user demand profiles should be further learned from the mobile data that local users generated.

Indeed, the centralization of baseband processing functionality poses great stresses and challenges on connections bridging the front-end RRHs and the back-end BBUs, due to the network capacity constraints, which will limit the performance of the overall system. To deal with the capacity constraints of such connections,

Bi et al. in [14] re-considered the scheme of computing resource allocation and proposed a hybrid computing structure to cope with this limited capacity problem mentioned above. Specifically, some computing tasks are proposed to remain at BSs to reduce the transmission burden to/from the cloud. Peng et al. in [19] proposed to utilize some high-power BSs as a fronthaul for control signal broadcasting, which not only reduces the transmission burden to/from the cloud but also mitigates the heterogeneous coordination problem between the C-RAN and the traditional cellular networks. Indeed, the tradeoff between the centralized and the distributed computing of radio access networks is still an open problem, together with the heterogeneous coordination between C-RAN and traditional cellular networks.

References

1. Q. Han, S. Liang, and H. Zhang, "Mobile cloud sensing, big data, and 5G networks make an intelligent and smart world," *IEEE Network*, vol. 29, no. 2, pp. 40–45, Mar. 2015.
2. Y. Cai, F. Yu, and S. Bu, "Cloud computing meets mobile wireless communications in next generation cellular networks," *IEEE Network*, vol. 28, no. 6, pp. 54–59, Nov. 2014.
3. Z. Sanaei, S. Abolfazli, A. Gani, and R. Buyya, "Heterogeneity in mobile cloud computing: Taxonomy and open challenges," *IEEE Communications Surveys & Tutorials*, vol. 16, no. 1, pp. 369–392, First 2014.
4. M. Satyanarayanan, P. Bahl, R. Caceres, and N. Davies, "The case for VM-based cloudlets in mobile computing," *IEEE Pervasive Computing*, vol. 8, no. 4, pp. 14–23, Oct. 2009.
5. A. Vulimiri, P. B. Godfrey, R. Mittal, J. Sherry, S. Ratnasamy, and S. Shenker, "Low latency via redundancy," in *Proceedings of the 9th ACM Conference on Emerging Networking Experiments and Technologies (CoNEXT)*, Santa Barbara, CA, Dec. 9–12, 2013, pp. 283–294.
6. B. Bangerter, S. Talwar, R. Arefi, and K. Stewart, "Networks and devices for the 5G era," *IEEE Communications Magazine*, vol. 52, no. 2, pp. 90–96, Feb. 2014.
7. A. Vulimiri, O. Michel, P. B. Godfrey, and S. Shenker, "More is less: Reducing latency via redundancy," in *Proceedings of the 11th ACM Workshop on Hot Topics in Networks*, Redmond, WA, Oct. 29–30, 2012, pp. 13–18.
8. Y. Cao, P. Hou, D. Brown, J. Wang, and S. Chen, "Distributed analytics and edge intelligence: Pervasive health monitoring at the era of fog computing," in *Proceedings of the 2015 Workshop on Mobile Big Data*, Hangzhou, China, Jun. 22–25, 2015, pp. 43–48.
9. M. Quwaider and Y. Jararweh, "Cloudlet-based efficient data collection in wireless body area networks," *Simulation Modelling Practice and Theory*, vol. 50, pp. 57–71, Jan. 2015.
10. G. Wang, T. Ng, and A. Shaikh, "Programming your network at run-time for big data applications," in *Proceedings of the 1st ACM workshop on Hot topics in Software Defined Networks (HotSDN)*, Helsinki, Finland, Aug. 13–17, 2012, pp. 103–108.
11. L. Cui, F. R. Yu, and Q. Yan, "When big data meets software-defined networking: SDN for big data and big data for SDN," *IEEE Network*, vol. 30, no. 1, pp. 58–65, Jan. 2016.
12. D. Kreutz, F. M. Ramos, P. Esteves Verissimo, C. Esteve Rothenberg, S. Azodolmolky, and S. Uhlig, "Software-defined networking: A comprehensive survey," *Proceedings of the IEEE*, vol. 103, no. 1, pp. 14–76, Jan. 2015.
13. S.-H. Park, O. Simeone, O. Sahin, and S. Shamai, "Joint precoding and multivariate backhaul compression for the downlink of cloud radio access networks," *IEEE Transactions on Signal Processing*, vol. 61, no. 22, pp. 5646–5658, Nov. 2013.
14. S. Bi, R. Zhang, Z. Ding, and S. Cui, "Wireless communications in the era of big data," *IEEE Communications Magazine*, vol. 53, no. 10, pp. 190–199, Aug. 2015.

15. H. Ahlehagh and S. Dey, "Video-aware scheduling and caching in the radio access network," *IEEE/ACM Transactions on Networking*, vol. 22, no. 5, pp. 1444–1462, Oct. 2014.
16. A. Liu and V. K. N. Lau, "Cache-enabled opportunistic cooperative MIMO for video streaming in wireless systems," *IEEE Transactions on Signal Processing*, vol. 62, no. 2, pp. 390–402, Jan. 2014.
17. M. Cha, H. Kwak, P. Rodriguez, Y.-Y. Ahn, and S. Moon, "I tube, you tube, everybody tubes: Analyzing the world's largest user generated content video system," in *Proceedings of the 7th ACM SIGCOMM Conference on Internet Measurement*, San Diego, CA, Oct. 24–26, 2007, pp. 1–14.
18. E. Bastug, M. Bennis, and M. Debbah, "Living on the edge: The role of proactive caching in 5G wireless networks," *IEEE Communications Magazine*, vol. 52, no. 8, pp. 82–89, Aug. 2014.
19. M. Peng, K. Zhang, J. Jiang, J. Wang, and W. Wang, "Energy-efficient resource assignment and power allocation in heterogeneous cloud radio access networks," *IEEE Transactions on Vehicular Technology*, vol. 64, no. 11, pp. 5275–5287, Nov. 2015.

Chapter 4
Computing

Mobile big data analytics demand high performance computing to accommodate the "5V" features and other distinct characteristics, in order to facilitate potential applications and services. In fact, sequential computing at a single machine cannot fulfill such computational demand over the tremendous amount of data, and a single machine may not even be able to hold the entire dataset in its memory, especially as the Moore's law is fading nowadays. Therefore, parallel computing over multiple nodes is of great importance in the era of (mobile) big data.

In this section, we provide an overview on existing computing solutions, which may be adopted for mobile big data analytics to better match the special character-istics of mobile big data (e.g., real-time). A complete (mobile) big data computing solution consists of two main components: the computing platform (hardware) and the computing architecture (software). In fact, a system of large-scale distributed commodity machines working parallelly in a distributed manner is a generally preferred computing platform in terms of flexibility and capital cost. Such a large-scale distributed computing platform has gained popularity along with the maturing software development of large-scale distributed computing [1, 2], which is deployed atop the hardware. In particular, the software is developed to efficiently utilize the large-scale distributed hardware resources, whose system architecture generally consists of three layers, namely the data injection layer, the data analytic layer, and the data storage layer. Each of them focuses on a distinct functionality in such a large-scale distributed computing system. In the rest of this section, we will first introduce the large-scale distributed computing platform (hardware), and then present the key properties required for large-scale distributed computing (software), as well as the other details of its system architecture.

© Springer International Publishing AG, part of Springer Nature 2018
X. Cheng et al., *Mobile Big Data*, Wireless Networks,
https://doi.org/10.1007/978-3-319-96116-3_4

4.1 Hardware

In the era of (mobile) big data, the fact that the complexity of big data processing exceeds the computing capability of a single node leads to massive parallel computing, which consists of multiple processors. In general, the very nature of a massive parallel computing system is determined by its two main components (as shown Fig. 4.1), namely the participating processors and the network architecture connecting those processors. Heterogeneous computing arises when processors of different types, e.g., central processing units (CPUs) and graphic processing units (GPUs), are involved in parallel computing. The networking among the participating processors, on the other hand, naturally plays a critical role in a system containing a massive number of processors in that it fundamentally determines the coupling level of nodes and in turn the capability of the system. In the sequel, we will discuss the computing platform for big data processing from these two perspectives.

4.1.1 Heterogeneous Computing

The legacy principle of a parallel system design is to interconnect the multiple similar processors into one system [3], which is termed as homogeneous computing. On the contrary, heterogeneous computing aims to exploit the strengths of different types of processors with intelligent load balancing to improve the computational performance [4].

   In heterogeneous computing, the specialized accelerators, e.g., GPUs or reconfigurable logics (FPGA), are integrated into machines with CPUs to accelerate the computing. The accelerators are connected to CPUs via an external bus, from

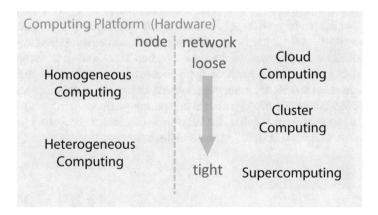

Fig. 4.1 Large-scale computing platform

which the heterogeneity arises. The commodity GPUs are the most commonly-used accelerators, which are originally designed for the game industry in image and graphic computing and later extended to support parallel data computing (termed as the general purpose GPU (GPGPU)). The power of GPUs lies in their thousands of cores, which could provide up to a few tera float operations per seconds (TFLOPS). In fact, the price/performance ratio is one of the most important advantages of the GPU-accelerated heterogeneous computing, compared with a cluster of nodes with similar computing capability without GPUs. For example, the NVIDIA 1080 graphic card could provide about 9 TFLOPS, which costs only a few hundred US dollars.

The CUDA package from NVIDIA provides the application programming interfaces (APIs) to utilize any CUDA-enabled GPUs for general computing [5]. The parallelism of CUDA-enabled GPU could be hierarchically divided into two levels: the GPU has multiprocessors (MPs); each MP has multiple stream processors (SPs). The SPs share a fast but small memory (typical 16 KB), based on which threads run in the SPs could synchronize their states. A global memory, larger but relatively slower (8 GB in a NVIDIA 1080 graphic card), is shared across the MPs. A collection of threads, termed as a block, is scheduled to feed a MP, in which the SPs are the ones that actually carry out the computations [6]. The parallelism in GPUs is accomplished by independent computations performed in the SPs. However, the performance gain of GPUs could be bottlenecked by the communication bandwidth between the GPUs and the CPUs [7]. Once the data or model size exceeds a GPU's global memory capacity, the frequent communications and data exchanges between the CPU and the GPUs would largely reduce the computing gain. Therefore, another level of parallelism and scalability has been developed to mitigate this problem, in which data or models are divided and allotted to a cluster of GPUs or a cluster of machines accelerated by GPUs [8–10].

4.1.2 Computing Systems

The actual operation of computing systems with a massive number of nodes heavily depends on the network connecting all these nodes. In the one extreme, multiple nodes in proximity are tightly interconnected with a dedicated local network, such that the system with multiple nodes could be regarded as a single virtual machine. This is the case with high performance computing (a.k.a., supercomputing). Supercomputing could date back to the 1960s, which was introduced by Seymour Cray [11]. The supercomputer usually consists of tens of thousands of processors in proximity, which is regarded as a single virtual machine. At the beginning, the architecture with customized vector processors and (distributed) shared memory is generally utilized in supercomputers, where the memories are accessed in a uniform addressing space. However, the architecture consisting of commodity processors and distributed memories has gained popularity in 2000s, where data exchange and sharing are fulfilled via message passing. In both cases, however, the interconnection of nodes in a supercomputer plays a critical role in the resultant computing performance. The interconnection in a supercomputer is generally accomplished by a customized dedicated high-speed local network, such as InfiniBand [12] (e.g., IBM Roadrunner) and Torus [11] (e.g., Cray XE6). The supercomputer originally designed for large volume data processing and computationally intensive simulations is also suitable for (mobile) big data analytics. However, the supercomputer with large-scale multiple processors tightly connected by a customized interconnection technology requires enormous capital investment, which is generally not available to big data researchers. In addition, the centralized administration of supercomputers leads to extra costs on the system operation and maintenance (e.g., cooling system).

In the other extreme, multiple nodes, possibly geographically distributed, are connected via, e.g., the Internet, such that a large computing task could be divided into multiple small tasks executed independently at each node. A well known example of this is the so-termed cloud computing. Cloud computing consists of multiple nodes that are often geographical distributed with computing and storage capability. These nodes are might be loosely connected by the Internet. The cloud computing is aimed to enable ubiquitous, convenient, and on-demand public computing services [13] hosted by vendors, such as Amazon Web Service (AWS), Microsoft Azure, and Google App Engine. Three service models, namely infrastructure as a service (IaaS), platform as a service (PaaS), and software as a service (SaaS), provide different levels of pay-as-your-go price models in terms of the amount of consumed computing resources. The business models of cloud computing are enabled by a virtualization layer between the physical hardware and the operating system at each node. The virtualization layer [14] is an abstraction of the underlying hardware including processors, memory, storage, and networks, which acts as a finer-grained computing resource unit, fulfilling the flexible on-demand computing requests. That is, a physical node could host several logical machines and the usage of the latter determines the actual cost of a particular computing task. However, to host the data at a public infrastructure may lead to privacy concerns, especially for the privacy-sensitive mobile big data. Details of big data analytics via cloud computing can be found in [15].

Lying in the middle ground between the two aforementioned extremes, cluster computing is an interconnected multi-node system without a customized high-speed local networks. Generally, a cluster may refer to a computing system that consists of multiple (tens to hundreds) commodity processors interconnected by general Ethernet networks. In the literature, this is termed as Beowulf cluster [11]. In addition, the cluster with multiple nodes working together closely could also be regarded as one single machine, which runs an non-proprietary OS. In fact, cluster computing could be sometimes regarded as a small-scale supercomputer [16, 17]. On the one hand, the utilization of commodity networking technology could largely reduce the capital investment, due to the economies-of-scale effect on general commercial hardwares. On the other hand, the limited communication capability of cluster computing leads to a bottleneck of performance, which greatly reduces the applicability for those which require large-scale frequent data exchange and sharing. Nevertheless, cluster computing is capable of dealing with task-independent applications. That is, a large task could be divided into small tasks that could be executed independently without frequent data exchange. Most of (mobile) big data problems could be formulated in this manner such that cluster computing can be applied.

4.2   Software

The large-scale computing software layer not only provides an easy-to-use interface for users to implement their data analytic algorithms as a single- or multi-thread program, but also facilitates utilization of the distributed computing resources provided by the hardware layer.

4.2.1   Key Properties and Architecture

In a large-scale computing, the parallelization of the programming or computing paradigm could be roughly categorized into two types, namely algorithmic parallelization and data parallelization [18]. The algorithmic parallelization essentially divides the algorithm into multiple tasks, which could be simultaneously processed at multiple computing nodes. The challenging issue of algorithmic parallelization is the communication among the processes at different nodes. The message passing interface (MPI) [19] is a typical standard on communications in distributed parallel computing among multiple nodes. In fact, the algorithmic parallelization is suitable for complex algorithms, which is usually applied to both streaming computational systems and data mining systems. On the other hand, data parallelization is more intuitive in the era of big data, as the data could be partitioned into multiple groups and each node in the system could process each data group independently while focusing on a single computing task. Hence, data parallelization is always adopted in the batch computational system. Nevertheless, all program models for scalable and parallel computing should satisfy some key properties as detailed below.

- **Scalability.** The scalability allows a large-scale computing system to utilize all distributed computing resources efficiently. In fact, a new computing architecture needs to be carefully designed in term of scalability to accommodate a tremendous amount of data by minimizing the communication requests from a large number of commodity machines (which could be scaled to tens of thousands of machines).
- **Fault Tolerance and Recovery.** Due to the massive number of nodes at the large-scale data center (tens of thousands of nodes), faults are inevitable. Faults may result from network congestions, disk errors, or scheduled offline maintenance. In fact, unreachable nodes are usually also recognized as faults. The computing

system should be able to tolerate such faults, by automatically rescheduling jobs at the faulted nodes to standby nodes and restoring the intermediate results from the failing nodes. In addition, a robust system design should facilitate easy recovery from human faults, as those are also inevitable.

- **Robustness to Stragglers.** Stragglers refer to the slow nodes in a system, which hold the entire computing process from proceeding. The sluggishness of computation at a node may be due to the competition of computing resources (e.g., CPU, memory, local disk) among multiple tasks. The program model should be able to deal with stragglers and reassign their jobs to alternative nodes.
- **Data Locality.** The major bottlenecks of a distributed computing system are the I/O of disks and the network bandwidth. Due to the massive data volume, the computing paradigm would transfer the computing task to nodes with the data rather than moving data across the network.

In the literature, there are many projects related to large-scale distributed computing. However, none of them can provide a one-size-fits-all solution for all (mobile) big data analytics problems. A real-time (mobile) big data analytics system should have three main components (shown in Fig. 4.2), namely the *data ingestion layer*, the *data analytic layer* and the *data storage layer* [20].

- **Data Ingestion Layer.** The data injection layer is aimed to absorb and queue the data from multiple sources, and to provide asynchronous communications, with a certain degree of fault tolerance, between the data source and data processing units. A typical open source data injection layer is the Apache Kafka [21], which manages data in a message-passing manner based on the publish and subscribe scheme.
- **Data Analytic Layer.** Inside the data analytic layer, there are also three components: the batch system, the stream system, and the data mining system [20]. Actually, the computation on the entire raw dataset is laid upon the batch computational system, due to its high throughput. However, batch computation is usually applied to the entire data set, resulting in a high latency. The output of the

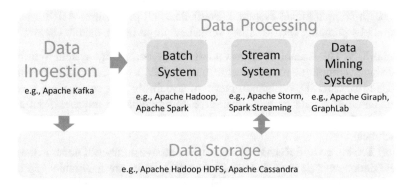

Fig. 4.2 Large scale distributed computing architecture

batch computational system is often outdated when it is ready. Hence, a stream computation system is developed to process recent data streams that the batch computational system cannot handle. The stream computational system could be of high throughput as well as low latency while dealing with much smaller data sizes. In addition, large-scale data mining is aimed to accommodate data mining or machine learning algorithms efficiently in the big data computing system.

A high-level architecture, Lambda Architecture [1], further details the data analytic layer. Besides the batch system and the stream system, a serve system is separated from the batch system to store outputs of batch computing (a.k.a, batch views in [1]). The results corresponding to the user query (a function on the whole set of data) will be computed based on both the batch views and the real-time views that are respective outputs of the batch computational system and stream computational system. In fact, the batch computational system will reproduce the batch views by re-computing the entire available dataset periodically rather than simply integrating the stream views with its outdated batch views. This will enhance the tolerance against the stream computational system faults or even the batch computational system faults. However, the robustness of such a re-computation scheme is at the cost of performance loss, as the batch system always needs to process the entire dataset.

- **Data Storage Layer.** To enhance the reliability of a big data system, the raw data stored in the big data system is designed to be eternal and immutable, such that any results could be recovered by re-computations on the immutable raw data. Hence, the data storage layer should provide a reliable storage service for the constantly growing master dataset. The scalability of a storage system is very much desired, since holding the massive volume of data as a single machine is neither practical nor reliable. In addition, the high-performance parallel reading should be supported for scalable batch computing in the data analytic layer. Distributed file systems are good candidates for the storage of raw data, for they provide fault tolerance and scalability [1]. The most commonly-used distributed file system is the Hadoop distributed file system (HDFS), which is a critical component of Apache Hadoop [22].

Next, we will discuss some widely adopted systems for the data analytic systems with more details.

4.2.2   Batch Computing Systems

The   concept of MapReduce, proposed by Google in 2004 [23], is a popular batch computing paradigm based on data parallelization. Although it has gained popularity in various applications, MapReduce has some inherent drawbacks such as high latency for iterative algorithms. To address such limitations, AMPLab of the University of California, Berkeley, proposed a new programming model, namely the resilient distributed dataset (RDD) in 2010 [24].

MapReduce

The MapReduce programming model is proposed to handle the massive data volume by a cluster of thousands of nodes in a parallel manner [23]. The open-source implementation of MapReduce is Apache Hadoop [22], which is a complete solution of distributed cluster computing based on MapReduce. MapReduce consists of two key tasks, map and reduce. The input data is first partitioned into multiple groups, which are then processed in parallel by map functions of independent nodes or workers (mappers). Intermediate results produced by mappers are stored locally (usually on disks) at each node. After all map tasks are completed, the intermediate results will be shuffled and fed to reduce tasks (reducers), which further process the intermediate results to final results as desired. The final results are saved at a global storage. A master node schedules computational jobs among workers and monitors their respective status (idle, in-progress or completed).

As MapReduce is designed to operate on multiple machines, the scalability issue is well resolved in MapReduce. In terms of fault tolerance, the master node in MapReduce will periodically ping every worker. Unreachable workers are regarded as failing nodes. The failed map tasks will be re-executed at other nodes, as intermediate results are stored at the local storage. The failed reduce tasks, however, may not need to be re-executed, as the final results are saved at the global storage. In the original implementation of MapRduce, a global file system (GFS) divides each file into multiple equi-length blocks, and keeps multiple copies (typically 3 copies) at the local storage. The master node generally assigns a map task to a node that contains a replica [25] to achieve data locality. When a MapReduce procedure is near completion, the master node will schedule redundant executions for tasks that are still in progress, in order to mitigate the negative effects caused by stragglers.

Resilient Distributed Dataset (RDD)

Due to the limitation of disk I/O, the in-memory computing with minimum data exchanges between the memory and the disk will significantly expedite the computation, especially when the in-memory working dataset will be reused in an iterative analysis algorithm. In MapReduce, each iteration would be expressed as a MapReduce job, which is inefficient [26] as the whole procedure involves lots of disk I/O operations. However, most mobile big data analytic algorithms will involve a great amount of iterative steps. In [27], Zaharia et al. proposed a distributed memory abstraction, termed as resilient distributed dataset (RDD), to support in-memory computation, while still maintaining MapReduce's attractive properties, such as fault tolerance, locality-aware scheduling, and scalability. The open-source implementation of RDD is Apache Spark [24].

An RDD is defined as a read-only partitioned collection of data, which could only be generated through deterministic operations (termed as transformation in Apache Spark) on either datasets in storage or other existing RDDs. After RDDs are defined, the computation will not be executed until an *action* is called, where the *action* is defined as an operation that returns values or exports data to a storage system. The partition of RDD is designed to support the property of scalability. The partition order could be controlled by users based on a key associated with each data record. In addition, the intermediate result caching could also be controlled and specified by users.

The reason why the RDD generation is restricted to deterministic operations on existing RDDs or datasets in stable storage, is to lower the overhead of the fault tolerance support. In general, two methods are employed to support fault tolerance: checkpointing (replicating the data) and data update logging (recording the differences). In fact, the replication of data across the network is expensive due to its massive volume; so is data update logging if many updates are made. Instead of recording the data itself, the RDD system will remember a series of deterministic operations (a.k.a., lineage [27]) that generate the current RDD. When an RDD partition is lost due to faults, the system will regenerate the current RDD based on the recorded lineage. In fact, the success of RDD relies on the sufficiency of the memory at each node, in comparison with MapReduce. When memories at a system are not sufficient, the performance gain of RDD could be marginal [28].

4.2.3   Stream Computing Systems

The distinct "real time" feature of mobile big data not only refers to the high velocity of the data stream, but also emphasizes the required responding speed of mobile big data based applications. In the literature, streaming computing is aimed to deal with such situations in large clusters (capable of cloud or cluster computing), but still maintaining the scalability, fault-tolerance, and locality properties.

Two Types of Stream Computational Systems

Unlike batch processing on the entire available dataset, the stream computing system is aimed to catch the recent data that the batch system cannot handle. The streaming computational system could be categorized into two types: one-at-a-time and micro-batched, respectively [1]. The one-at-a-time streaming computational system incrementally processes each data record in a predefined data stream one after another following some continuous models. The scalability of the one-at-a-time system lies in that the processing on each data record could be performed in parallel across multiple nodes in the system, i.e., algorithmic parallelization. One-at-a-time stream processing projects include Apache S4 [29], Apache Storm [30], and TimeStream [31]. In one-at-a-time systems, every data record will possibly trigger a new update on the real-time view by flowing through a predefined computing logic network, whose topology could be characterized by a directed acyclic graph (DAG). The DAG computational topology adopted in the one-at-a-time system will also simplify the design and implementation of algorithmic parallelization.

On the other hand, the micro-batched system first processes a small batch of data records at a time following a discretized model and then updates the real-time view by integrating the output of the current micro batch processing and the past real-time view. An example micro-batched system is D-Stream (Spark Streaming) [32], which utilizes the RDD computing paradigm in each micro batch process. Evidently, the one-at-a-time system has much lower latency, as it does not have to wait for a preset time interval. In addition, it takes more time to process a set of data records than just a single record, while the processing time is relatively much smaller than that of the micro-batched processing. Nevertheless, the micro-batched system has a higher throughput and is more robust than the one-at-a-time system, in which the fault and straggler could be recovered easily and rapidly. A survey and benchmark of existing streaming computational systems could be found in [33, 34].

Fault-Tolerant Methods

Two fault-tolerant methods are adopted in the one-at-a-time systems : the hardware replication and the upstream backup [32]. The hardware replication introduces redundancy to the system, by simultaneously keeping two sets of nodes. The upstream backup is to replay the entire computation on the backup data from the very beginning (e.g., Apache S4) or from the intermediate result (e.g., Apache Storm). The re-computation on the data from the very beginning in Apache S4 leads to very high fault recovery latency. The re-computation in Apache Storm is not general and is only applicable to some special operators [31]. In TimeStream, the system keeps monitoring the dependencies of the output and the state at each node in the computational network, and the lost output of the failed node is re-computed from its immediate saved states. However, the fault-tolerant method in batch processing, such as lineage in RDD, could be extended to each small batch computing task, as the real-time views are updated at the batch level and small batch processing tasks are independent of each other, which will enhance the fault tolerance.

4.2.4   Data Mining Systems

In mobile big data, the data from mobile users may not only be treated independently as in batch or stream processing, but also be jointly investigated in terms of their relative relationship or dependency. For example, the location of mobile users is studied [35] along with their social relationship. In general, the relationship among the data could be characterized by graphs (directed or undirected). In fact, the dependencies in data limit the applicability of MapReduce, since MapReduce requires strict data parallelism. The MapReduce-based iterative processing is significantly inefficient, due to the lack of direct iterative mechanisms in MapReduce. In addition, many machine learning or data mining algorithms update their parameters iteratively, e.g., training of deep belief networks. Therefore, efficient distributed computing in such scenarios is of a significant demand in mobile big data processing. Pregel [36] and GraphLab [37] are two typical graph-parallel abstractions in the literature. GraphX [38] implements both Pregel and GraphLab abstractions with a graph extension of RDD in Spark, termed as resilient distributed graph (RDG).

Pregel and Its Derivative

The pregel computing system is initially aimed to efficiently deploy large-scale graph computing in a cluster of commodity computers efficiently, which is built by Google [36] and its open-source version is maintained in Apache Giraph [39].

Similar to the MapReduce model, the graph is partitioned in terms of the vertices of the input graph and assigned to each node of the cluster. The parallelization of Pregel is bulk synchronous parallel (BSP). That is, Pregel synchronously carries out simultaneous computation on all vertices of the graph at each iteration of a graph algorithm, where each iteration is termed as a "superstep" in Pregel. The interaction of vertices in a graph algorithm is fulfilled by messages passed by other vertices of the graph in the previous iteration. In addition, messages pushed by vertices in one superstep could only be processed in the next superstep. In other words, the computing on the entire graph is synchronized at the granularity of supersteps and messages shuffle across the entire cluster before a new superstep begins. Such global synchronism makes the graph algorithm implementation easier and free of deadlocks or data races, but leads to high latency in general. The message passing model in Pregel improves the efficiency compared to MapReduce, since it does not need to update the state of the entire graph at each iteration as the chained MapReduce does. The fault-tolerance in Pregel is ensured by checkpointing, where nodes periodically store the states of each vertex at a global storage and the entire graph will be re-computed from the latest check point when one or more nodes fail.

The message staleness issue will arise when a message could not be seen and processed immediately but needs to wait until the next superstep. This will compromise the system performance due to node blocking for strict global synchronism on the entire graph. Giraph++ proposed in [40] aims to mitigate such an issue by allowing message exchanges among internal vertices within a subgraph to be immediately seen and processed. Hence, the computing of a graph moves from a vertex-centric manner to a graph-centric manner. GiraphUC proposed in [41] further reduces the communications and node blocking overhead due to global synchronization across the entire graph, by splitting a global superstep into multiple logical supersteps. The logical superstep is locally synchronized, which could reduce a large portion of communication blockings and stragglers in the system due to global synchronization. Such a programming model is termed as barrierless asynchronous parallel (BAP). In fact, Pregel and Giraph++ could both be regarded as some BAP special cases where a global superstep is divided into multiple logical supersteps. However, such hybrid models may lead to high coordination complexity in the system and compromise the ease of use for parallel graph algorithm implementation.

GraphLab

The GraphLab was originally developed for parallel graph computing, machine learning and data mining at a single machine with multiple cores [42]. It was later adapted to cluster computing in [37, 43]. GraphLab is also a vertex-centric abstraction. Unlike Pregel, GraphLab completely eliminates the synchronism requirement in message passing. Instead, Graphlab applies the shared memory to achieve asynchronism, which will in turn accelerate the convergence of a machine learning algorithm. The shared memory allows the vertex to read and write the state of each vertex following a pull model. The dynamic schedule of GraphLab, in which not all parameters are updated in the universal and uniform frequency could further reduce the number of iterations required for convergence. This is achieved by decoupling the computation schedule and the data movement in the system. However, asynchronism may lead to data inconsistency due to data races. Hence, data lock is demanded in GraphLab to ensure correctness. In GraphLab, different distributed lock models could be employed to solve this issue. However, distributed locking is resource-expensive and may lead to performance degradation. In addition, the termination of a computation task needs to be checked by a distributed consensus algorithm, due to the lack of global synchronization in GraphLab. The significant recent advancement in big data computing makes mobile big data processing possible with a cluster of commodity computers. However, the computing system design for a specific mobile big data task varies in terms of the data volume, data velocity, and specific requirements of the task. In general, all the subsystems described previously might be combined based on their advantages and the needs of specific mobile big data applications.

References

1. N. Marz and J. Warren, *Big Data: Principles and Best Practices of Scalable Realtime Data Systems*, 1st ed. Greenwich, CT, USA: Manning Publications Co., 2015.
2. D. A. Reed and J. Dongarra, "Exascale computing and big data," *Communication of ACM*, vol. 58, no. 7, pp. 56–68, Jun. 2015.

3. Y. Gao and P. Zhang, "A survey of homogeneous and heterogeneous system architectures in high performance computing," in *Proceeding of IEEE International Conference on Smart Cloud (SmartCloud)*, Nov 2016, pp. 170–175.
4. S. Mittal and J. S. Vetter, "A survey of cpu-gpu heterogeneous computing techniques," *ACM Computing Survey*, vol. 47, no. 4, Jul. 2015.
5. M. Harris, "Many-core gpu computing with nvidia cuda," in *Proceedings of the 22nd Annual International Conference on Supercomputing*, Island of Kos, Greece, Jun. 7–12, 2008, pp. 1–1.
6. R. Raina, A. Madhavan, and A. Y. Ng, "Large-scale deep unsupervised learning using graphics processors," in *Proceedings of the 26th Annual International Conference on Machine Learning*, ser. ICML '09, Montreal, Quebec, Canada, Jun. 14–17, 2009, pp. 873–880.
7. J. Dean, G. Corrado, R. Monga, K. Chen, M. Devin, M. Mao, M. aurelio Ranzato, A. Senior, P. Tucker, K. Yang, Q. V. Le, and A. Y. Ng, "Large scale distributed deep networks," in *Advances in Neural Information Processing Systems 25*, F. Pereira, C. J. C. Burges, L. Bottou, and K. Q. Weinberger, Eds. Curran Associates, Inc., 2012, pp. 1223–1231.
8. Z. Fan, F. Qiu, A. Kaufman, and S. Yoakum-Stover, "Gpu cluster for high performance computing," in *Proceedings of the 2004 ACM/IEEE Conference on Supercomputing*, Pittsburgh, PA, Nov. 2004, pp. 47–47.
9. V. V. Kindratenko, J. J. Enos, G. Shi, M. T. Showerman, G. W. Arnold, J. E. Stone, J. C. Phillips, and W. m. Hwu, "Gpu clusters for high-performance computing," in *Proceedings of IEEE International Conference on Cluster Computing*, New Orleans, LA, USA, Aug. 31 – Sep. 4, 2009, pp. 1–8.
10. A. Coats, B. Huval, T. Wang, D. J. Wu, and A. Y. Ng, "Deep learning with COTS HPC systems," in *Proceedings of the 30 th International Conference on Machine Learning*, Atlanta, Georgia, Jun. 16–21, 2013.
11. S. Limet, W. W. Smari, and L. Spalazzi, "High-performance computing: to boldly go where no human has gone before," *Concurrency and Computation: Practice and Experience*, vol. 27, no. 13, pp. 3145–3165, 2015.
12. G. F. Pfister, "An introduction to the infiniband architecture," in *High Performance Mass Storage and Parallel I/O*, J. Fagerberg, D. C. Mowery, and R. R. Nelson, Eds., 2001, ch. 42, pp. 617–632.
13. P. M. Mell and T. Grance, "Sp 800-145. the nist definition of cloud computing," Gaithersburg, MD, United States, Tech. Rep., 2011.
14. M. Armbrust, A. Fox, R. Griffith, A. D. Joseph, R. Katz, A. Konwinski, G. Lee, D. Patterson, A. Rabkin, I. Stoica, and M. Zaharia, "A view of cloud computing," *Communications of the ACM*, vol. 53, no. 4, pp. 50–58, Apr. 2010.
15. I. A. T. Hashem, I. Yaqoob, N. B. Anuar, S. Mokhtar, A. Gani, and S. U. Khan, "The rise of big data on cloud computing: Review and open research issues," *Information Systems*, vol. 47, pp. 98–115, Jan. 2015.
16. I. Foster, Y. Zhao, I. Raicu, and S. Lu, "Cloud computing and grid computing 360-degree compared," in *Proceedings of Grid Computing Environments Workshop*, Austin, TX, Nov. 16, 2008, pp. 1–10.
17. N. Sadashiv and S. M. D. Kumar, "Cluster, grid and cloud computing: A detailed comparison," in *Proceeding of the 6th International Conference on Computer Science Education (ICCSE)*, Singapore, Aug. 3–5, 2011, pp. 477–482.
18. H. Mohamed and S. Marchand-Maillet, "Distributed media indexing based on MPI and MapReduce," *Multimedia Tools and Applications*, vol. 69, no. 2, pp. 513–537, Nov. 2014.
19. E. Gabriel, G. E. Fagg, G. Bosilca, T. Angskun, J. J. Dongarra, J. M. Squyres, V. Sahay, P. Kambadur, B. Barrett, A. Lumsdaine, R. H. Castain, D. J. Daniel, R. L. Graham, and T. S. Woodall, *Open MPI: Goals, Concept, and Design of a Next Generation MPI Implementation*. Berlin, Heidelberg: Springer Berlin Heidelberg, 2004, pp. 97–104.
20. R. Ranjan, "Streaming big data processing in datacenter clouds," *IEEE Cloud Computing*, vol. 1, no. 1, pp. 78–83, May 2014.
21. "Apache Kafka," 2016. [Online]. Available: http://kafka.apache.org/
22. "Apache Hadoop," 2016. [Online]. Available: http://hadoop.apache.org/

23. J. Dean and S. Ghemawat, "Mapreduce: Simplified data processing on large clusters," in *Proceedings of the 6th Conference on Symposium on Operating Systems Design & Implementation*, vol. 6, San Francisco, CA, Dec. 5, 2004, pp. 137–149.
24. "Apache Spark," 2016. [Online]. Available: http://spark.apache.org/
25. J. Dean and S. Ghemawat, "Mapreduce: Simplified data processing on large clusters," *Commun. ACM*, vol. 51, no. 1, pp. 107–113, Jan. 2008.
26. M. Zaharia, M. Chowdhury, M. J. Franklin, S. Shenker, and I. Stoica, "Spark: Cluster computing with working sets," in *Proceedings of the 2nd USENIX Conference on Hot Topics in Cloud Computing*, Boston, MA, Jun. 22–25, 2010, pp. 1–7.
27. M. Zaharia, M. Chowdhury, T. Das, A. Dave, J. Ma, M. McCauley, M. J. Franklin, S. Shenker, and I. Stoica, "Resilient distributed datasets: A fault-tolerant abstraction for in-memory cluster computing," in *Proceedings of the 9th USENIX Conference on Networked Systems Design and Implementation*, San Jose, CA, Apr. 25–27, 2012, pp. 1–15.
28. L. Gu and H. Li, "Memory or time: Performance evaluation for iterative operation on Hadoop and Spark," in *Proceedings of the 10th IEEE International Conference on High Performance Computing and Communications & IEEE International Conference on Embedded and Ubiquitous Computing (HPCC_EUC)*, Zhangjiajie, China, Nov. 13–15, 2013, pp. 721–727.
29. L. Neumeyer, B. Robbins, A. Nair, and A. Kesari, "S4: Distributed stream computing platform," in *Proceedings of IEEE International Conference on Data Mining Workshops*, Sydney, Australia, Dec. 13, 2010, pp. 170–177.
30. "Apache Storm," 2016. [Online]. Available: http://storm.apache.org/
31. Z. Qian, Y. He, C. Su, Z. Wu, H. Zhu, T. Zhang, L. Zhou, Y. Yu, and Z. Zhang, "Timestream: Reliable stream computation in the cloud," in *Proceedings of the 8th ACM European Conference on Computer Systems*, Prague, Czech Republic, Apr. 14–17, 2013, pp. 1–14.
32. M. Zaharia, T. Das, H. Li, T. Hunter, S. Shenker, and I. Stoica, "Discretized streams: Fault-tolerant streaming computation at scale," in *Proceedings of the Twenty-Fourth ACM Symposium on Operating Systems Principles*, Farmington, Pennsylvania, Nov. 3–6, 2013, pp. 423–438.
33. X. Liu, N. Iftikhar, and X. Xie, "Survey of real-time processing systems for big data," in *Proceedings of the 18th International Database Engineering & Applications Symposium*, Porto, Portugal, Jul. 7–9, 2014, pp. 356–361.
34. S. Qian, G. Wu, J. Huang, and T. Das, "Benchmarking modern distributed streaming platforms," in *Proceedings of IEEE International Conference on Industrial Technology (ICIT)*, Taipei, Taiwan, Mar. 14–17, 2016, pp. 592–598.
35. E. Cho, S. A. Myers, and J. Leskovec, "Friendship and mobility: user movement in location-based social networks," in *Proceedings of the 17th ACM SIGKDD International Conference on Knowledge Discovery and Data Mining*, San Diego, CA, Aug. 21–24, 2011, pp. 1082–1090.
36. G. Malewicz, M. H. Austern, A. J. Bik, J. C. Dehnert, I. Horn, N. Leiser, and G. Czajkowski, "Pregel: A system for large-scale graph processing," in *Proceedings of the 2010 ACM SIGMOD International Conference on Management of Data*, Indianapolis, Indiana, USA, Jun. 6–10, 2010, pp. 135–146.
37. Y. Low, D. Bickson, J. Gonzalez, C. Guestrin, A. Kyrola, and J. M. Hellerstein, "Distributed graphlab: A framework for machine learning and data mining in the cloud," *Proceedings of VLDB Endowment*, vol. 5, no. 8, pp. 716–727, Apr. 2012.
38. R. S. Xin, J. E. Gonzalez, M. J. Franklin, and I. Stoica, "Graphx: A resilient distributed graph system on spark," in *Proceedings of First International Workshop on Graph Data Management Experiences and Systems*, New York City, NY, Jun. 23, 2013, pp. 2:1–2:6.
39. "Apache Giraph," 2016. [Online]. Available: http://giraph.apache.org/
40. Y. Tian, A. Balmin, S. A. Corsten, S. Tatikonda, and J. McPherson, "From "think like a vertex" to "think like a graph"," *Proceedings of VLDB Endowment*, vol. 7, no. 3, pp. 193–204, Nov. 2013.

41. M. Han and K. Daudjee, "Giraph unchained: Barrierless asynchronous parallel execution in pregel-like graph processing systems," *Proceedings of VLDB Endowment*, vol. 8, no. 9, pp. 950–961, May 2015.
42. Y. Low, J. Gonzalez, A. Kyrola, D. Bickson, C. Guestrin, and J. M. Hellerstein, "Graphlab: A new framework for parallel machine learning," *Proceedings of Conference on Uncertainty in Artificial Intelligence (UAI)*, Jul. 8–11, 2010.
43. J. E. Gonzalez, Y. Low, H. Gu, D. Bickson, and C. Guestrin, "Powergraph: Distributed graph-parallel computation on natural graphs," in *Proceedings of the 10th USENIX Conference on Operating Systems Design and Implementation*, Hollywood, CA, Oct. 8–10, 2012, pp. 17–30.

Chapter 5
Applications

5.1 Overview

Analytics and mining over mobile big data enriched with time and location information will provide great opportunities for new services. The potential applications driven by such mobile user data could be roughly divided into two categories. One is mining on the individual user data to provide personalized services (e.g., context-aware sensing, point of interests, activity recognition, etc.). The other is mining on the aggregation of mobile user data to learn and analyze the pattern of human activities, which aims to understand human behaviors in order to help public service planning and city management (e.g., social response monitoring in lieu of social events or disasters, anomaly detection, traffic flow pattern learning, city zone characterization, etc.). However, these two categories are not strictly separable, as some services not only rely on unique patterns mined from individual user data, but also depend on common patterns analyzed from the aggregation of mobile data from multiple users. Details of several typical applications exploiting mobile user data are discussed as follows (Fig. 5.1).

5.1.1 Mobility

The human mobility is of great interest in sociology but has not been thoroughly studied due to the lack of fine-grained geolocation data to record trajectories of individual human beings. The location sequence data collected in the mobile big data can facilitate not only the study and analysis on the human behavior at both individual and aggregated scales, but also mobility prediction based on the behavior pattern revealed by the spatiotemporal dynamics embedded in the data.

© Springer International Publishing AG, part of Springer Nature 2018
X. Cheng et al., *Mobile Big Data*, Wireless Networks,
https://doi.org/10.1007/978-3-319-96116-3_5

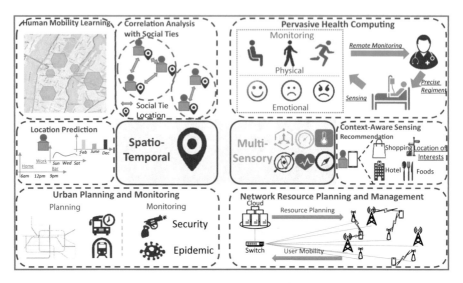

Fig. 5.1 Applications of mobile big data

Fundamental Analysis on Human Mobility

The study and analysis of human mobility is of great interest, especially with the availability of the tremendous volume of spatio-temporal mobile data directly recording people's daily life. As suggested in [1], according to the study on 100,000 anonymized mobile phone users whose positions were tracked for a 6-month period, it was observed that human mobility is highly regularized rather than randomized in both temporal and spatial domains. Additionally, each individual followed a reproducible pattern in terms of characteristic travel distance (exploration) and a significant probability of returning to highly frequented locations (preferential return). The inherent similarity of individual travel patterns has great potentials in public applications driven by human mobilities, such as urban planning, epidemic prevention, and emergency response.

Furthermore, Song et al. [2] aimed to quantify the role of randomness in human behavior to answer the question that to what extent the human mobility can be predicted, based on the cell IDs in mobile CDR of 50,000 individuals each visiting more than two locations. With entropy modeling on human behavior randomness, e.g., the number of locations visited, the heterogeneous visiting probabilities of different locations, and the time spent at each location, the predictability of individual mobility could be quantified. As a result, it was concluded in [2] that the human mobility is predictable with up to 93% accuracy based on the user entropy empirically determined from the mobile data and the Fano's inequality.

On the other hand, the privacy protection issue is under significant challenge in the context of mobile big data with spatio-temporal information, as it is shown that the trajectory of an individual is unique and can be easily recognized with 90%

accuracy only based on four spatio-temporal points [3], regardless of the anonymity in the spatio-temporal mobile data. Intuitively, as the spatio-temporal resolution is getting coarse, the distinct trajectories also become less identifiable. However, this study showed that when the resolution decreases, the trajectory identifiability decays at a rate that is a magnitude of order slower. In other words, even with low-resolution data, user trajectories could still be identified with a relatively high distinction.

Location Prediction Over Different Time Scales

With semantically meaningful location information, user location prediction is an interesting application of mobile big data. Such interest is further enhanced by the aforementioned high predictability of human mobility we discussed early. Based on these, location prediction has been studied for short [4], medium [5], and long terms [6], respectively. We expect that the full exploitation of location prediction based on human mobility will potentially sprout a great variety of new applications. For each temporal scale of location prediction, the details are discussed as follows.

The *short-term* prediction aims to forecast the destination of users within an hour or less. In [4], Krumm et al. proposed a framework called predestination, which attempts to predict the destination based on the GPS trajectory history as the trip progresses over short terms. The destination of a user is recognized based on the length of their sojourn time. In [4], the prediction results were not limited to locations visited in the past but also include the unvisited places, based on the concept of an open-world model with explicit characterization of the model incompleteness. Prediction was carried out by the probabilistic analysis of user destinations and the spatial modeling of a given region. The prediction results could reach median errors of about 2 km based on 3667 different GPS trajectories. Instead of predicting the destination, Ziebart et al. [7] presented a probabilistic model based on the Markov decision process to characterize observed user behaviors, in order to predict the turns and routes of users during a trip within several minutes.

The *medium-term* mobility prediction intends to first analyze the mobility pattern of human beings on a daily basis and then predict the behavior of an individual within a day or so. Eagle et al. [5] modeled the daily behavior of an individual by the weighted aggregation of eigenvectors termed as *eigenbehavior*, a set of principal components generated by the covariance matrix of behavior data collected in the Reality Mining study. However, only four locations were considered in [5] for simplicity, namely home, work, elsewhere, and no signal. In fact, the principal component analysis on an individual behavior reveals that the daily behavior of an individual can be characterized at up to 90% accuracy with only six principal components. Furthermore, with the principal components learned from the personal dataset, one can predict the behaviors for the rest of the day based on the weights learned from a set of half-day data. In addition, individuals can be clustered into communities based on the similarity of their eigenbehaviors within the society.

Sadilek et al. [6] proposed a nonparametric method to extract the significant and effective patterns in human mobility, aiming to predict one's location in the *long-term* future (on the order of months or years). The authors first applied Fourier analysis to recognize significant periodicities in human mobility and then extracted the significant patterns based on the principal component analysis (PCA), on a 32,268-day worth of GPS location dataset, where the latitude and longitude are represented by a complex number (real and imaginary parts, respectively). In addition, the method was shown to work with very coarse location information, where triangular grids at a 400 m resolution are used. The accuracy of the long-term location prediction was claimed to be 80% for up to 80 weeks as shown in [6].

Correlation Analysis Between Social Tie and Human Mobility

Intuitively, the human mobility is influenced by social ties. As a result, the physical location and social relationships are intrinsically entangled [8–11]. Actually, exploitation of correlation between human mobility and social ties is expected to bring twofold benefits: more regularized location prediction for individuals or social groups with improved accuracy and precision; enhanced identification of social ties and interest groups that in turn facilitates better targeted services.

In [9], a model was introduced by explicitly considering the feedback of mobility on the formation of social ties using data from Twitter, Gowalla, and Brightkite. A model integrating mobility and social interactions has the potential of revealing characteristics of the network topology in terms of total number of the connected components, the distance distribution among connected users, the properties of social overlaps, and the distance-based user clusters. In fact, it was shown in [10] that the exploitation of trajectory correlation between two individuals can improve the human mobility prediction dramatically, where the correlation is characterized by their mutual information.

The geographical distance and its impact on network topologies were mainly studied in [9]. However, the distance between two continuous trajectories may not clearly capture the social tie between them, as multiple different routes in an urban area could reach the same location. Therefore, Toole et al. in [11] defined the mobility as discrete visits to places at different time stamps and proposed explicit measures on the patterns of both mobility and social behaviors, in order to quantify the correlation between two individuals, based on which the mobility similarity could be used for semantic social relationship classification, e.g., as coworkers, family members, etc. The mobility similarity between two people is quantified based on their mobility traces of specific times on a given day, which essentially amounts to the calculation of the angle between two vectors representing the CDR-recorded trajectories of two users. In addition, the study also reveals that the temporal correlation inherent in user trajectories plays an important role in the classification of social relationships. For example, the mobility similarity during daytime on weekdays strongly indicates the relationship of coworkers. It also suggests that human mobility is far more similar to each other when two people have certain social relationships than when the two are strangers.

Context-Aware Sensing and Recommendation

Context-aware recommendation is an immediate application of mobile big data embedded with spatio-temporal information. It aims at providing precise personal recommendation to users, given the context of a user (i.e., location, time stamp, velocity, acceleration, etc.), as well as the user's personal preferences, where the location information is the fundamental information for context-aware services. Context-aware recommendation is essentially the interaction between the user preferences acquired from his/her behavior history and the available options under a certain context in terms of time, space, environment etc. [12]. Not surprisingly, nearly all context-aware services require certain location information. The integration of the actual location, the inferred points of interest, the predicted location based on personal location and/or social relationships, and the inferred/refined social roles based on location information, as we discussed before, will undoubtedly enhance the consumer experience of context-aware recommendations. In addition, the location prediction also plays an important role in context-aware recommendations [13].

The common preference among multiple users is important for precise recommendation, as the users in the same group given a specific context have a high probability of sharing some common interests. Furthermore, to mine the context-aware preferences from the context-rich mobile big data is challenging, as the data from one user or a limited number of users might not be sufficient for accurate preference learning. In [14], it was proposed to first derive common context-aware preferences from the data of multiple users, and then to characterise individual user preferences via a probabilistic distribution of these common preferences.

In addition, the social roles of users could be studied based on the abstracted characterization of shared preferences or behavior patterns among multiple users [15], which could be utilized for recommendation or advertisement based on similar social roles. The benefits of social-role based recommendation is more precise and more reliable, as the cross recommendation among users with the same social role can be applied due to the social nature of human beings. In fact, the common preferences among users are not the only factor that influences the choice of a user. In [16], the distance (how far between a candidate restaurant and the user) combined with the common preferences (recorded by the ranks of candidate restaurants) was considered to estimate the likelihood that a user will choose a particular candidate restaurant nearby.

5.1.2 Pervasive Health Computing

With the multi-sensor data from wearable devices or smart phones, mobile health (mHealth) is a promising application. Such data from sensors can passively and unobtrusively collect the necessary information for health monitoring. In addition, smart phones with rich connectivity can provide a platform for active data collection in many scenarios, such as facial expression capturing with phone cameras and patient audio clip recording by microphones.

Actually, physical activity monitoring is an intuitive application of health monitoring [17] with such a rich set of sensors. For instance, fall detection coupled with an alert system [18] could be implemented to detect the fall and alert the authorities at the same time. In [19], Wu et al. collected the data of the accelerometer and the gyroscope for 16 participants on 13 activities, such as sitting, walking, jogging, and going upstairs and downstairs. Overall, the results of monitoring are claimed to achieve very good accuracies: 52.3–79.4% for up-and-down-stair walking, 91.7% for jogging, 90.1–94.1% for walking on the ground, and 100% for sitting.

In fact, mobile sensors equipped in smart devices can help doctors monitor their patients remotely and frequently. The sensing results will further help doctors customize the personalized medical treatment for each patient. In [20], Sharma et al. proposed a framework to help with monitoring Parkinson patients via both active and passive approaches, to understand the daily activities and manage the complex medication regimens personalized to individual needs. In particular, passive monitoring relies on accelerometers as well as gyroscopes to monitor motor aspects of Parkinson patients, such as walking, falls, balance, tremor, and so on, while active monitoring relies on the collection of contextual data, such as speech, facial tremors, etc., by interacting with patients.

Although physical activity monitoring can be considered as a modernized integration of biomedical sensors into personal mobile communication devices, the mental health monitoring hinges more upon context sensing and the development of emotion learning [21]. Based on where we have been, with whom we communicate, what applications we use, and how we use our mobile devices, various learning algorithms can be developed to exploit these mobile phone sensor values. These learning models can be adapted to predict a mobile user's moods, emotions, cognitive/motivational states, activities, environmental contexts, and social contexts.

In [22], LiKamWa et al. proposed a mood monitor based on the logged data collected from 32 participants over 2 months. The authors showed that by analyzing the communication history and application usage patterns, a user's daily mood curve could be statistically inferred with an initial accuracy of 66%, which gradually improves to an accuracy of 93% after a 2-month personalized training period. High-quality cameras equipped in smart phones [21] could easily capture our facial expressions indicating emotions and moods, and the new development on emotion learning techniques will further facilitate mobile emotion monitoring. In [23], the smile intensity mined from facial expressions was studied to help machines understand the emotion of human beings, by tracking the changes of facial muscles leading to a specific expression.

Besides psychological monitoring and care for patients, such information can also be employed for a vast array of health/mood related applications. For example, video and music recommendation, context-aware advertisement, and personalized social networking would all significantly benefit from these.

5.1.3 Public Services

Besides personal applications and services stated above, the aggregation of mobile big data could provide a great tool to represent the big picture of human being social behaviors, e.g., human mobility, or social response and propagation of events, diseases, and disasters. In other words, the mobile big data can help with understanding how the individual dynamics shape the structure of cities, which in turn assists better urban planning and public service planning.

For instance, mobile big data can help reveal regions of different functionalities in urban areas [24, 25] in terms of the aggregation patterns of human mobility. In [24], Grauwin et al. discovered a universal structure of cities in terms of general functionalities, based on the analysis of CDRs in three major cities. Actually, the fine-grained functionalities of a region in the city may be overlapping. In [25], Yuan et al. segmented a city into disjointed regions according to major roads and represent each region by a probabilistic distribution on a set of functions, which are learned via a latent variable model (topic-based inference model) with two 3-month GPS trajectory datasets.

Based on the urban study, urban planning (e.g., public transportation planning [26]) can be effectively designed. Traffic patterns can be inferred and different traffic zones can be determined based on the mobile big data, such as CDRs [27, 28]. In [27], Dong et al. first semantically extracted and tagged the origins and destinations of traffic commuters from the CDR and utilize mobile phones as well as base stations as sensors to measure the traffic trend in the city. The distribution of travel times learned from mobile device sensors (e.g., accelerators) and GPS trajectories can help monitor the traffic condition and provide the fastest route information for each individual in the area based on the current traffic status [28]. In addition, the social and pathological response and propagation of events, diseases, and disasters can be studied based on the current trends of social networks, physical and emotional sensing, and call data records, which helps improve public security, as well as emergency response and recovery [29, 30].

In fact, urban planning roots in the understanding over the aggregate human mobility in an urban area. The most fundamental problem for urban planning or even network resource management (discussed in next subsection) is the crowd flow analysis in a city. The crowd flow monitoring and prediction could provide a solid foundation for urban sensing and planning. Recent work [31] on crowd flow prediction utilizes the recent development of neural networks (deep learning), which will be discussed in Sect. 5.2. Overall, the crowd flow analysis based on mobile big data is still an open problem.

5.1.4   Network Planning and Management

Naturally, based on the aforementioned features, mobile big data provides excellent prior information to feed the resource allocation and optimization algorithms that are driving the communication networks. In particular, mobile big data can be used to extract, model, and predict the patterns of mobile traffic, which has great potentials towards more efficient network planning and resource management. With innovative development of the cloud radio access network (C-RAN), centralized management of multiple radio access points may be capable of adapting the allocation of network resources according to the dynamics of mobile users, by jointly exploiting the mobile big data that is collected by system operators simultaneously over multiple points [32].

Mobile big data can also be used to gain individual insights such as demographic attributes, mobility patterns, personal preferences, and instant context. Such insights can be used to optimize personal content delivery, contextual services, and mobile advertisement. These knowledge not only can improve the network planning from the traditional perspective in terms of access point planning and flexible radio resource management, but also will be essential for emerging applications such as adaptive content distribution by providing content consumption cartography [33]. In fact, the specific problem formulation on the data-driven communication network resource planning, management, and scheduling is still an open issue, which requires the expertise from both the data mining and communication communities.

5.2   Methodology

As stated previously, machine learning and data mining (MLDM) tools are critical for mobile big data applications. In fact, MLDM is an interdisciplinary area, crossing statistics, mathematics, computing etc., which are traditionally applied to speech recognition, computer vision, natural language processing and so on. In this subsection, we will review some important concepts of machine learning on mobile big data applications and discuss some future trends in the context of mobile big data.

5.2.1 Representation

Generally, the raw data cannot be directly fed to a learning machine, as it may contain redundancy, noise, outliers, etc., which will negatively impact the learning performance. Hence, the data should be first preprocessed prior to the learning process. Actually, the data cleaning alone may not be sufficient, as the raw data sometimes may not even render the features needed for a particular application. Hence, data representation is vital in each machine learning application. This is termed as feature engineering. In fact, a large portion of endeavors in machine learning are devoted to the feature engineering.

For example, locations in mobile big data are the most common yet very critical information. The longitude and latitude information from GPS locations are the utmost raw data, which are samples of the continuous geospace. The CDR contains the discrete locations of the base stations and records such locations corresponding to a Cell ID. The GPS information could be either used directly in its original form without any further processing, as in [6, 24, 36], or processed to provide more informative or convenient features. For example, the number of distinct locations as well as the sojourn time spent at a specific location could be derived from the raw GPS data [2]. On one hand, good features extracted from the raw data could enhance the learning performance of learners, thus making information extraction easier in predictors or capturing the posterior distribution of the underlying explanatory more precisely in probabilistic modeling. On the other hand, if not executed appropriately, feature extraction may lead to detail compromise and information loss from the original data [34]. Hence, feature engineering is indeed a fine art on its own.

Though important, feature engineering is quite labor-intensive in the development of MLDM applications. The future trend is to make feature engineering more automatic rather than manual. This could be fulfilled by a special type of learning [35], namely "representation learning." Representation learning enables MLDM to largely bypass manual feature engineering based on human intuition and prior knowledge, especially under the context of variety inherited in big data. The representation learning has been studied and applied in speech recognition, object recognition, and natural language processing. Due to the distinct multi-sensory and spatial-temporal characteristics in mobile big data, it may be overwhelming to extract features manually from such high-dimensional data. Hence, representation learning, may come handy in helping formulate a good representation in order to achieve good MLDM performance on mobile big data.

5.2.2 Models

MLDM models could be categorized into two types, namely the descriptive and the predictive. The descriptive type is aimed to illustrate the dependency or relationship among the data, while the predictive type estimates these based on the models trained by labeled data. Depending on whether the data is labeled, the categorization into descriptive and predictive types could also be termed as unsupervised and supervised learning, respectively. The specific models of MLDM used in mobile big data applications are discussed as follows.

Descriptive Methods (Unsupervised)

- Clustering or Segmentation. Clustering is utilized to categorize data into several groups based on the similarities within the group. For example, the type of land use in a city (city functionalities) is segmented using a clustering technique (k-means) on the CDR data in [24].
- Principal Component Analysis (PCA). The PCA is originated from eigen analysis in matrix theory. The PCA is typically applied to learn significant features while reducing the noise or error nested in the data. In [5], daily human mobility behaviors are represented by behavior features, in which eigenvectors (eigen-behaviors) are generated by PCA from the discrete location data in CDR. Then, the daily human mobility behavior is represented as weighted superimposition of the eigenbehaviors, which could be applied in mobility prediction.
- Probabilistic Graphical Model (PGM). This model can be used to characterize probabilistic dependencies among different features or even among entries of the raw data with some prior knowledge or assumptions. The captured dependencies could be characterized by graphs (directed or undirected). For example, generative probabilistic models are originally applied to learn latent topics underlying the text corpora, where words in the corpora are assumed to be generated according to a probability distribution of latent topics. Along with similar assumptions, such a topic model is applied to learn city functionalities in [25], in which popularity density reviewed by the mobile big data is assumed to be generated by city functionalities.

Predictive Methods (Supervised)

- Regression and Classification. Regression is aimed to build a function based on the training dataset to generate continuous valued outputs; whereas, classification utilizes the labeled historical dataset to build a classifier, to predict a categorized output given a set of features. The classifier could be linear (e.g., linear discriminant analysis) or non-linear (e.g., support vector machine (SVM)). In the mobile big data context, activity recognition or context-aware related applications based on multi-sensory data from smart phones [18, 19] are typical classification problems.

Besides these two types of categories, "reinforcement learning" has also been proposed and developed in the past years, for which the training data comes from the interaction between the learning machine and the feedback of its environment

[36]. In fact, thousands of MLDM models have been presented in the literature and more are published every year. However, to choose an appropriate model that could successfully characterize a specific problem with mobile big data needs significant efforts. In fact, three critical components in a machine learning application should be carefully investigated in this new context: representation, evaluation and optimization [37].

In addition, other special challenges pertinent to MLDM should also be studied. One of the challenges in MLDM is the overfitting problem. The overfitting problem arises when the model (generally a predictor or classifier) characterize too much noise from the training dataset rather than capturing the real merits in data. This will lead to poor performance in validation tests, but could be potentially mitigated by exploiting the large size of data set in mobile big data [37]. The overfitting problem is also critical in the neural network, an important tool that mimics how human brain organizes neurons and learns the environments. Until 2006, the neural network research has been stuck with the overfitting and the high dimensional training issues. "Deep learning" [38, 39] was proposed as a neural network in 2006 with multiple hidden layers (and thus deep). Pretraining via unsupervised learning on each layer of a neural network, deep learning has been successfully applied to several fields. The success of deep learning roots in the progress of computing capabilities as well as the tremendous size of training dataset. In fact, such trends may also be extended to mobile big data analytics. It has been demonstrated in [40] that the deep neural network could achieve better performance than other MLDM techniques (e.g., random forests) in context-aware activity recognition.

Furthermore, unsupervised learning based on the deep neural network structure is attracting significant attention, e.g., autoencoder [39] or generative adversarial networks [41], which could learn the inner feature of data without labels. Such new techniques may potentially facilitate many novel applications, which remains as an active research direction.

5.2.3   Knowledge Discovery

Nearly every application or service of mobile big data will be based on the mapping of the raw data to some useful information, and then from the discovered information to intelligent decision. In such, one may be prone to the traps inherent in the complicated knowledge extraction from raw mobile big data, which will in turn lead to false knowledge discovery. The prevention from false knowledge discovery requires deep understanding of application, along with careful verification and validation [42].

Moreover, the conflict between the real-time response requirement of mobile big data applications and extremely large data volumes coupled with velocities not only places great challenges on the infrastructure supporting these applications, but also poses challenges to the algorithm design in mobile big data mining [43]. Such algorithms should be scalable and adaptable to dynamic mobile user environments.

In addition, how to extract and select features from the multi-sensor mobile big data according to the specific application/service requirement is of great importance, while maintaining reasonable computation complexity in data collection and data processing.

In summary, mobile big data with unique temporal-spatio features will have unprecedented potentials in a great variety of applications for both personal and public services.

5.3 User Modeling

Most of the aforementioned applications of mobile big data are pertinent to personalized customization of either individual users or user groups. This inevitably necessitates the understanding of user interests to specific information, their behavior traits, and user tendencies during a given time period. Such concepts of user profiling can be dated back to [44–48], which refer to the acquisition of each user's basic information, interest and behavior, together with the establishment of a description file [49]. Such user modeling can be regarded as a data preprocessing procedure in order to extract important features from mobile big data to profile each mobile user, which could be further fed to learning machines and data miners to facilitate personalized applications and services.

Although the objective of user modeling is to facilitate precise and dynamic personalized services, description of user interests is not the whole story. Lying in the core of a personalized information service system, computability is another basic requirement in user modeling. In other words, user modeling is an algorithmic and structured user description, which abstracts and extracts computable user models from user interests and behaviors, such as browsed content, browsing behavior, surfing context, and geographic location. In addition, the derived models should be able to evolve with user interests that may vary dynamically. Next, we will overview mobile user modeling from the perspective of the data source. We will see that with different sources, the type, format and application of the data also vary dramatically.

5.3.1 With Data from OTT Servers

With OTT servers, we could directly capture the user browsing information. With user interests and preferences being the key concerns of content and service providers, it is natural to develop user models based on user browsing behaviors readily available at OTT servers, such that content/service recommendations can be provided to users and new content services can be developed.

The raw information at the OTT servers includes texts, user profiles, system logs, and URLs, etc. Due to the large quantity of such data, it is not practical to perform detailed analysis on each record, and pre-filtering and pre-processing are often necessary to form somewhat compressed records in order to facilitate the computability of the ensuing data processing and analytics. In particular, the raw HTTP inquiry records are usually filtered to expunge information that are not closely related to the core accessed content. Examples of typically expunged information include graphics, animations, and scripts. The OTT server then combines the filtered records from the same IP address during a short time period to form an approximation of the actual user browsing behavior. In the mean time, as the actual usefulness of the URL access data in indicating user preferences lies in its represented information type, the Open Directory Project (ODP) was devised [50], which is a web directory of Internet resources hierarchically arranged by subject and provides an open framework establishing an ontology of URLs. Using this tool, the user browsing record can be interpreted as the corresponding website type, whose statistics can then lead to user preference profiling.

One of the most important information that can be obtained from the OTT server data is the interaction between users and webpages, i.e., the access rate of each webpage and its corresponding user sojourn time. With this aspect, the mobile user modeling based on the OTT server data is essentially web usage mining [51–53], in which the interactions between users and webpages are analyzed and modeled in order to infer user behaviors and preference models. As the main purpose of OTT servers in user modeling is content and service recommendations, which are more user-oriented and personalized, user modeling here is often jointly operated with recommendation systems.

The overall procedure consists of two aspects: data preparation and conversion vs. analysis and modeling. In the data preparation and conversion process, one needs to merge and abstract the data obtained from the servers. One basic abstraction of contents or services is to abstract the raw server data into a pageview, which typically contains the ID (usually a URL), Type (service provided, e.g., information vs. merchandise), and Content (keywords, subject, etc.). One basic user abstraction is to abstract a user visit into a so-termed "Session". Following data preparation and conversion is analysis and modeling, for which some currently adopted methods are listed below:

- **Clustering**. In data preparation, one essentially obtains the interactions among users and contents. Hence the classification methods are either user-oriented or content-oriented. Most commonly adopted algorithms include K-Means [54], hierarchical clustering [55], and expectation-maximization (EM) [56].
- **Association Rule Discovery [57–60]**. In [61], clustering and association rule discovery methods are combined, where the user browsing behaviors are first clustered using Relational Fuzzy Subtractive Clustering (RFSC), and the recommendation results are then obtained by applying association rule discovery within each cluster.
- **Latent Variable Models (LVMs)**. Statistical models are used to reveal the underlying relations and structures embedded in the data. Most commonly adopted models are Factor Analysis (FA) models [62] and Finite Mixture Models (FMM) [63].

From the above, it is evident that, as content and service providers, the OTT servers can conveniently establish user interest and preference models from the captured user browsing information, to address their key concerns of personalized content/service recommendations, renovations, and creations. However, user models established solely based on the OTT server data lack the direct information of the user locations, device/app utilization habits, and call/text information, etc. This renders joint processing over data from multiple sources were meaningful and necessary.

5.3.2   *With Data from Mobile Devices*

Compared with the data from OTT servers, the device system log and sensor information usually lead to more intimate information about the user. Proper interpretation and analysis of these raw data can reveal the user app utilization habits, device utilization routines, background environments and even user physical and psychological status. Several example applications along this line have been introduced in Sect. 5.1. The typical user modeling procedure often contains the following steps:

- **Feature Construction and Extraction**. User feature vectors are constructed from the features extracted from the raw data. Evidently, the selection of features is critically important here, and could vary significantly in different scenarios. For example, [64] considered global and partial user features; [65] examined the inclusion of device utilization behaviors and user locations selected as features, whereas cross validation was adopted in [66] for feature selection.
- **Data Cleaning**. Frequently, missing data needs to be added and feature vectors need to be regularized, leading to dimension reduction. For example, F-Test, Relief, Principal Component Analysis (PCA) and kernel PCA were discussed and applied in [64].
- **Model Establishment**. In [64], the Support Vector Machine (SVM) classifier was employed, and a decision tree was used to fuse results obtained from different models in [66].
- **Prediction and Recommendation**. In this step, user statistical characteristics based on gender, age, marital status, employment, household size, and app utilization habits can all be incorporated into the base of prediction and recommendation [64, 66, 67].

Although data from mobile devices often directly reveal authentic and personal user information, retrieval of such data necessitates access to the device system log and sensor information, which are either privacy-sensitive or only available to certain mobile software developers/providers. Hence, aggregation of such data from a diversity of terminal applications, a user pool of different types and a variety of devices may be challenging.

5.3.3   With Data from Network Operators

In communications networks, all user network access behaviors are recorded, leading to a comprehensive user pool, diverse user behaviors, and multidimensional user data. In addition, network operators could further analyze the signaling messages from the communication network gateway to acquire customer numbers, terminal IDs, communication behaviors, texting behaviors, surfing times, browsed contents, operating systems, access point names, traffic flows, and precise locations.

All these facilitate in-depth user modeling involving usage preference profiling, online routine prediction, location tracking, and device characterization, just to name a few. Compared to the data from OTT servers, the data from network operators allow for reconstruction of the physical settings of the user access, such as the corresponding time, location and context. Compared to the data from mobile devices, the data from network operators are from all users across the network, from different types of devices, and for various terminal applications, making large-scale data mining and statistical modeling possible. With such data, it is possible to not only make content/service recommendations based on user interest modeling alone as with the OTT server data, but also fine-tune such recommendations based on user mobility traces or activity routines. If further coupled with the user mood analysis from call/text records and device/app utilization, the recommendations could be further adapted to the user psychological status.

As a result, mining the mobile user data from network operators could reveal not only what the users want, but also where, when, and how the users want. The main research issue is then how to make customized and optimized selection of user features for specific applications. For example, in [68], user features were constructed based on user behavior, utilization, location characterization, etc., and K-means classification was then applied to enable effective router selection by administrators. A number of research results also utilized the location information together with the context of user relationship to infer the user trajectory [69–71], or user interests [71, 72]. The user trajectories, together with traffic data, can facilitate realistic network traffic flow simulations of the entire network [72], which may even assist transportation administration and optimization [71]. In some research efforts, the communication record is used to infer the user interests and groups. For instance, communication logs and feedbacks at the customer service centers were used to help operators construct better customer satisfaction models in [73]. In [74], communication behaviors and relationships among users were used to infer the features of their neighbors in the network.

To summarize, in terms of both the variety and accessibility, data from network operators provides the greatest potential for next-generation precision personalized content/merchandise services, thanks to the comprehensiveness, abundance, multi-dimensionality, and continuity in both space and time. In the longer term, its integration with the more detailed content information at the OTT servers and more intimate user information at mobile devices, will open new doors towards fine-grained mobile service personalization. To bring all these together, an interweaving supporting infrastructure will become a necessity, as we will discuss in next sections. The need of new paradigms supporting large-scale mobile data collection, processing and sharing brings not only challenges but also great research opportunities, which will be discussed in the next section.

References

1. M. C. Gonzalez, C. A. Hidalgo, and A.-L. Barabasi, "Understanding individual human mobility patterns," *Nature*, vol. 453, no. 7196, pp. 779–782, Mar. 2008.
2. C. Song, Z. Qu, N. Blumm, and A.-L. Barabási, "Limits of predictability in human mobility," *Science*, vol. 327, no. 5968, pp. 1018–1021, Feb. 2010.
3. Y.-A. de Montjoye, C. A. Hidalgo, M. Verleysen, and V. D. Blondel, "Unique in the crowd: The privacy bounds of human mobility," *Scientific Reports*, vol. 3, Mar. 2013.
4. J. Krumm and E. Horvitz, "Predestination: Inferring destinations from partial trajectories," in *Proceedings of the 8th International Conference on Ubiquitous Computing*, Orange County, CA, Sep. 17–21, 2006, pp. 243–260.
5. N. Eagle and A. Pentland, "Eigenbehaviors: identifying structure in routine," *Behavioral Ecology and Sociobiology*, vol. 63, no. 7, pp. 1057–1066, May 2009.
6. A. Sadilek and J. Krumm, "Far out: Predicting long-term human mobility." in *Proceedings of the 26th AAAI Conference on Artificial Intelligence*, Toronto, Canada, Jul. 22–26, 2012.
7. B. D. Ziebart, A. L. Maas, A. K. Dey, and J. A. Bagnell, "Navigate like a cabbie: Probabilistic reasoning from observed context-aware behavior," in *Proceedings of the 10th International Conference on Ubiquitous Computing*, Seoul, Korea, Sep. 21–24, 2008, pp. 322–331.
8. E. Cho, S. A. Myers, and J. Leskovec, "Friendship and mobility: user movement in location-based social networks," in *Proceedings of the 17th ACM SIGKDD International Conference on Knowledge Discovery and Data Mining*, San Diego, CA, Aug. 21–24, 2011, pp. 1082–1090.
9. P. A. Grabowicz, J. J. Ramasco, B. Gonalves, and V. M. Eguiluz, "Entangling mobility and interactions in social media," *PLoS ONE*, vol. 9, no. 3, p. e92196, Mar. 2014.
10. M. De Domenico, A. Lima, and M. Musolesi, "Interdependence and predictability of human mobility and social interactions," *Pervasive and Mobile Computing*, vol. 9, no. 6, pp. 798–807, Dec. 2013.
11. J. L. Toole, C. Herrera-Yaqüe, C. M. Schneider, and M. C. González, "Coupling human mobility and social ties," *Journal of The Royal Society Interface*, vol. 12, no. 105, pp. 1–9, Feb. 2015.
12. J. Zhuang, T. Mei, S. C. Hoi, Y.-Q. Xu, and S. Li, "When recommendation meets mobile: Contextual and personalized recommendation on the go," in *Proceedings of the 13th International Conference on Ubiquitous Computing*, Beijing, China, Sep. 17–21, 2011, pp. 153–162.
13. B. Shaw, J. Shea, S. Sinha, and A. Hogue, "Learning to rank for spatiotemporal search," in *Proceedings of the 6th ACM International Conference on Web Search and Data Mining*, Rome, Italy, Feb. 4–8, 2013, pp. 717–726.
14. H. Zhu, E. Chen, H. Xiong, K. Yu, H. Cao, and J. Tian, "Mining mobile user preferences for personalized context-aware recommendation," *ACM Transactions on Intelligent Systems and Technology*, vol. 5, no. 4, pp. 58:1–58:27, Dec. 2014.
15. R. K. Wong, V. W. Chu, and T. Hao, "Online role mining for context-aware mobile service recommendation," *Personal Ubiquitous Computing*, vol. 18, no. 5, pp. 1029–1046, Jun. 2014.
16. R. Kumar, M. Mahdian, B. Pang, A. Tomkins, and S. Vassilvitskii, "Driven by food: Modeling geographic choice," in *Proceedings of the 8th ACM International Conference on Web Search and Data Mining (WSDM)*, Shanghai, China, Jan. 1–Feb. 6, 2015, pp. 213–222.

17. Y. Cao, P. Hou, D. Brown, J. Wang, and S. Chen, "Distributed analytics and edge intelligence: Pervasive health monitoring at the era of fog computing," in *Proceedings of the 2015 Workshop on Mobile Big Data*, Hangzhou, China, Jun. 22–25, 2015, pp. 43–48.

18. F. Sposaro and G. Tyson, "iFall: An android application for fall monitoring and response," in *Proceedings of the Annual International Conference of Engineering in Medicine and Biology Society (EMBC)*, Minneapolis, MN, Sep. 3–6, 2009, pp. 6119–6122.

19. W. Wu, S. Dasgupta, E. E. Ramirez, C. Peterson, and J. G. Norman, "Classification accuracies of physical activities using smartphone motion sensors," *Journal of Medical Internet Research*, vol. 14, no. 5, p. e130, Oct. 2012.

20. V. Sharma, K. Mankodiya, F. De La Torre, A. Zhang, N. Ryan, T. G. Ton, R. Gandhi, and S. Jain, "SPARK: Personalized parkinson disease interventions through synergy between a smartphone and a smartwatch," in *Design, User Experience, and Usability. User Experience Design for Everyday Life Applications and Services*, A. Marcus, Ed. Grete, Greece: Springer, 2014, pp. 103–114.

21. A. Moore, "Why 2016 could be a watershed year for emotional intelligence–in machines," 2016. [Online]. Available: http://blogs.scientificamerican.com/guest-blog/why-2016-could-be-a-watershed-year-for-emotional-intelligence-in-machines/

22. R. LiKamWa, Y. Liu, N. D. Lane, and L. Zhong, "Moodscope: Building a mood sensor from smartphone usage patterns," in *Proceedings of the 11th Annual International Conference on Mobile Systems, Applications, and Services*, Taipei, Taiwan, Jun. 25–28, 2013, pp. 389–402.

23. J. M. Girard, J. F. Cohn, and F. De la Torre, "Estimating smile intensity: A better way," *Pattern Recognition Letters*, vol. 66, pp. 13–21, Nov. 2015.

24. S. Grauwin, S. Sobolevsky, S. Moritz, I. Gdor, and C. Ratti, "Towards a comparative science of cities: Using mobile traffic records in New York, London, and Hong Kong," in *Computational Approaches for Urban Environments*, ser. Geotechnologies and the Environment, M. Helbich, J. Jokar Arsanjani, and M. Leitner, Eds. Springer International Publishing, Nov. 2015, vol. 13, pp. 363–387.

25. J. Yuan, Y. Zheng, and X. Xie, "Discovering regions of different functions in a city using human mobility and POIs," in *Proceedings of the 18th ACM SIGKDD International Conference on Knowledge Discovery and Data Mining*, Beijing, China, Aug. 12–16, 2012, pp. 186–194.

26. M. Berlingerio, F. Calabrese, G. Lorenzo, R. Nair, F. Pinelli, and M. L. Sbodio, "AllAboard: A system for exploring urban mobility and optimizing public transport using cellphone data," in *Proceedings of European Conference on Machine Learning and Knowledge Discovery in Databases*, Prague, Czech Republic, Sep. 23–27, 2013, pp. 663–666.

27. H. Dong, M. Wu, X. Ding, L. Chu, L. Jia, Y. Qin, and X. Zhou, "Traffic zone division based on big data from mobile phone base stations," *Transportation Research Part C: Emerging Technologies*, vol. 58, pp. 278–291, Sep. 2015.

28. J. Yuan, Y. Zheng, C. Zhang, W. Xie, X. Xie, G. Sun, and Y. Huang, "T-drive: Driving directions based on taxi trajectories," in *Proceedings of the 18th SIGSPATIAL International Conference on Advances in Geographic Information Systems*, San Jose, CA, Nov. 2–5, 2010, pp. 99–108.

29. F. Antonelli, M. Azzi, M. Balduini, P. Ciuccarelli, E. D. Valle, and R. Larcher, "City sensing: Visualizing mobile and social data about a city scale event," in *Proceedings of International Working Conference on Advanced Visual Interfaces*, Como, Italy, May 27–30, 2014, pp. 337–338.

30. B. Moumni, V. Frias-Martinez, and E. Frias-Martinez, "Characterizing social response to urban earthquakes using cellphone network data: The 2012 Oaxaca earthquake," in *Proceedings of the 2013 ACM Conference on Pervasive and Ubiquitous Computing Adjunct Publication*, Zurich, Switzerland, Sep. 8–12, 2013, pp. 1199–1208.

31. J. Zhang, Y. Zheng, D. Qi, R. Li, and X. Yi, "Dnn-based prediction model for spatio-temporal data," in *Proceedings of the 24th ACM SIGSPATIAL International Conference on Advances in Geographic Information Systems*, Burlingame, California, Oct. 31 – Nov. 3, 2016, pp. 92:1–92:4.

32. S. Bi, R. Zhang, Z. Ding, and S. Cui, "Wireless communications in the era of big data," *IEEE Communications Magazine*, vol. 53, no. 10, pp. 190–199, Aug. 2015.

33. S. Hoteit, S. Secci, Z. He, C. Ziemlicki, Z. Smoreda, C. Ratti, and G. Pujolle, "Content consumption cartography of the Paris urban region using cellular probe data," in *Proceedings of the 1st Workshop on Urban Networking*, Nice, France, Dec. 10–13, 2012, pp. 43–48.
34. G. James, D. Witten, T. Hastie, and R. Tibshirani, *An introduction to statistical learning*. Springer, 2013, vol. 6.
35. Y. Bengio, A. Courville, and P. Vincent, "Representation learning: A review and new perspectives," *IEEE Transactions on Pattern Analysis and Machine Intelligence*, vol. 35, no. 8, pp. 1798–1828, Aug 2013.
36. L. P. Kaelbling, M. L. Littman, and A. W. Moore, "Reinforcement learning: A survey," *Journal of artificial intelligence research*, vol. 4, pp. 237–285, 1996.
37. P. Domingos, "A few useful things to know about machine learning," *Commun. ACM*, vol. 55, no. 10, pp. 78–87, Oct. 2012.
38. G. E. Hinton and R. R. Salakhutdinov, "Reducing the dimensionality of data with neural networks," *Science*, vol. 313, no. 5786, pp. 504–507, Jul. 2006.
39. I. G. Y. Bengio and A. Courville, "Deep learning," 2016, book in preparation for MIT Press. [Online]. Available: http://www.deeplearningbook.org
40. M. A. Alsheikh, D. Niyato, S. Lin, H. p. Tan, and Z. Han, "Mobile big data analytics using deep learning and apache spark," *IEEE Network*, vol. 30, no. 3, pp. 22–29, May 2016.
41. I. Goodfellow, J. Pouget-Abadie, M. Mirza, B. Xu, D. Warde-Farley, S. Ozair, A. Courville, and Y. Bengio, "Generative adversarial nets," in *Advances in Neural Information Processing Systems 27*, Z. Ghahramani, M. Welling, C. Cortes, N. D. Lawrence, and K. Q. Weinberger, Eds., 2014, pp. 2672–2680.
42. X. Wu, X. Zhu, G.-Q. Wu, and W. Ding, "Data mining with big data," *IEEE Transactions on Knowledge and Data Engineering*, vol. 26, no. 1, pp. 97–107, Jan. 2014.
43. N. Stojanovic, L. Stojanovic, Y. Xu, and B. Stajic, "Mobile CEP in real-time big data processing: Challenges and opportunities," in *Proceedings of the 8th ACM International Conference on Distributed Event-Based Systems*, Mumbai, India, May 26–29, 2014, pp. 256–265.
44. M. Agosti and M. Melucci, "Information retrieval on the web," in *Proceedings of the 3rd European Summer-School on Lectures on Information Retrieval*, Varenna, Italy, Sep. 11–15, 2000, pp. 242–285.
45. C. Buckley, A. Singhal, and M. Mitra, "Using query zoning and correlation within SMART," in *Proceedings of the 5th Text Retrieval Conference*, Gaithersburg, Maryland, Nov. 1996, pp. 105–118.
46. G. Salton, "The smart retrieval system," in *Experiments in Automatic Document Processing*. Prentice Hall, 1971.
47. L. Shastri, "Why semantic networks," in *Principles of Semantic Networks: Explorations in the Representation of Knowledge*. Morgan Kaufmann, 1991.
48. S. Thomas, "Http essentials: protocols for secure, scaleable web sites," in *Scalable Web Sites*. John Wiley, New York, Mar. 2001.
49. Y. H. Wu, Y. C. Chen, and L. P. Chen, "Enabling personalized recommendation on the web based on user interests and behaviors," in *Proceedings of the 11th International Workshop on Research Issues in Data Engineering*, Heidelberg, Germany, 2001, pp. 17–24.
50. "Open directory project." [Online]. Available: https://www.dmoz.org/
51. R. Cooley, B. Mobasher, and J. Srivastava, "Web mining: information and pattern discovery on the World Wide Web," in *Proceedings of the 9th IEEE International Conference on Tools with Artificial Intelligence*, Newport Beach, CA, Nov. 3–8, 1997, pp. 558–567.
52. B. Mobasher, "Web usage mining," *Web data mining: Exploring hyperlinks, contents and usage data*, vol. 12, pp. 1216–1220, 2005.
53. J. Srivastava, R. Cooley, M. Deshpande, and P. Tan, "Web usage mining: discovery and applications of usage patterns from web data," *ACM SIGKDD Explorations Newsletter*, vol. 1, no. 2, pp. 12–23, Jan. 2000.
54. L. Ungar and D. Foster, "Clustering methods for collaborative filtering," in *Proceedings of the AAAI Workshop on Recommendation Systems*, Madison, WI, Jul. 26–27, 1998.

55. A. Kohrs, A. Kohrs, B. Merialdo, and B. Merialdo, "Clustering for collaborative filtering applications," in *Computational Intelligence for Modelling, Control & Automation. Intelligent Image Processing, Data Analysis & Information Retrieval*, M. Mohammadian, Ed. IOS Press, 1999, pp. 199–204.

56. A. P. Dempster, N. M. Laird, and D. B. Rubin, "Maximum likelihood from incomplete data via the EM algorithm," *Journal of the Royal Statistical Society, Series B*, vol. 39, no. 1, pp. 1–38, Jan. 1977.

57. X. Fu, J. Budzik, and K. J. Hammond, "Mining navigation history for recommendation," New Orleans, LA, Jan. 9–12, 2000, pp. 106–112.

58. W. Lin, S. A. Alvarez, and C. Ruiz, "Efficient adaptive-support association rule mining for recommender systems," *Data Mining and Knowledge Discovery*, vol. 6, no. 1, pp. 83–105, Jan. 2002.

59. B. Mobasher, H. Dai, T. Luo, and M. Nakagawa, "Effective personalization based on association rule discovery from web usage data," in *Proceedings of the 3rd International Workshop on Web Information and Data Management*, Atlanta, Georgia, Nov. 5–10, 2001, pp. 9–15.

60. B. Sarwar, G. Karypis, J. Konstan, and J. Riedl, "Analysis of recommender algorithms for e-commerce," in *Proceedings of the 2nd ACM Conference on Electronic Commerce*, Minneapolis, MN, Oct. 17–20, 2000, pp. 158–167.

61. B. Suryavanshi, N. Shiri, and S. Mudur, "Improving the effectiveness of model based recommender systems for highly sparse and noisy web usage data," in *Proceedings of the IEEE/WIC/ACM International Conference on Web Intelligence (WI)*, Compiegne, France, Sep. 19–22, 2005, pp. 618–621.

62. Y. Zhou, X. Jin, and B. Mobasher, "A recommendation model based on latent principle factors in web navigation data," in *Proceedings of the 3rd International Workshop on Web Dynamics in conjunction with the 13th International World Wide Web Conference*, New York City, NY, May 18, 2004, pp. 52–61.

63. I. Cadez, P. Smyth, E. Ip, and H. Mannila, "Predictive profiles for transaction data using finite mixture models," Information and Computer Science Department, University of California, Irvine, Irvine, CA, Tech. Rep. 01–67, Dec. 2001.

64. M. Kaixiang, T. Ben, Z. Erheng, and Y. Qiang, "Report of Task 3: Your Phone Understands You," in *Proceedings of the Workshop on Nokia Mobile Data Challenge*, Newcastle, UK, Jun. 18–19, 2012.

65. W. M. C. Nadeem S, "Demographic prediction of mobile user from phone usage," in *Proceedings of the Workshop on Nokia Mobile Data Challenge*, Newcastle, UK, Jun. 18–19, 2012, pp. 16–21.

66. J. Ying, Y. Chang, C. Huang, and V. Tseng, "Demographic prediction based on users mobile behaviors," in *Proceedings of the Workshop on Nokia Mobile Data Challenge*, Newcastle, UK, Jun. 18–19, 2012.

67. J. Wang, C. Zeng, C. He, L. Hong, L. Zhou, R. Wong, and J. Tian, "Context-aware role mining for mobile service recommendation," in *Proceedings of the 27th Annual ACM Symposium on Applied Computing*, Trento, Italy, Mar. 26–30 2012, pp. 173–178.

68. L. Qian, B. Wu, R. Zhang, W. Zhang, and M. Luo, "Characterization of 3G data-plane traffic and application towards centralized control and management for software defined networking," in *Proceedings of IEEE International Congress on Big Data*, Santa Clara, CA, Jun. 27–Jul. 2, 2013, pp. 278–285.

69. L. Wang, K. Hu, T. Ku, X. Yan, and L. Wang, "Mining frequent trajectory pattern based on vague space partition," *Knowledge-Based Systems*, vol. 50, pp. 100–111, Sep. 2013.

70. X. Wu, K. N. Brown, and C. J. Sreenan, "Analysis of smartphone user mobility traces for opportunistic data collection in wireless sensor networks," *Pervasive and Mobile Computing*, vol. 9, no. 6, pp. 881–891, Dec. 2013.

71. Z. Sun and X. J. Ban, "Vehicle classification using GPS data," *Transportation Research Part C: Emerging Technologies*, vol. 37, pp. 102–117, Dec. 2013.

72. D. J. Yang W, Cheng H, "A location-aware recommender system for mobile shopping environments," *Expert Systems with Applications*, vol. 34, no. 1, pp. 437–445, Jan. 2008.
73. W. Hsu, G. Jacobsen, Y. Jin, and A. Skudlark, "Using social media data to understand mobile customer experience and behavior," in *Proceedings of the 22nd European Regional Conference of the International Telecommunications Society*, Budapest, Hungary, Sep. 18–21, 2011.
74. C. Herrera-Yagüe and P. J. Zufiria, "Prediction of telephone user attributes based on network neighborhood information," in *Proceedings of the 8th International Conference on Machine Learning and Data Mining in Pattern Recognition*, Berlin, Germany, Jul. 13–20, 2012, pp. 645–659.

Chapter 6
Case Study: Demand Forecasting for Predictive Network Managements

6.1 Background

6.1.1 Data-Driven Predictive Network Management

The mobile big data collected by mobile network operators can also benefit the management of mobile networks as stated previously. In fact, mobile big data could help uncover and understand user' behavior patterns [1] via effective data mining techniques, which could benefit to the resource-constraint network optimization, from network planning, network traffic monitoring to network management. In recent years, self-organizing networks (SON) is widely studied to automatically manage and organize networks without manual interventions [2, 3]. One motivation to employ SONs in cellular networks is the reduction of network operational expenditures (OPEX) and capital expenditures (CAPEX), which requires full exploits of the capability of network infrastructure. In fact, data plays a critical role in the cellular SONs, providing system observability and predictability for network management. In [4], three categories of data-driven SON functions in cellular networks are defined, namely self configuration, self optimization, and self healing, in which the self optimization can adapt network configurations to meet mobile demands in a real-time manner or in a predictive manner.

The demand forecasting will play an important role in various cellular SON functions to predictively optimize the network [5]. With the virtualization and cloudization of network functions in future cellular networks [6, 7], the physical networks will be sliced and operated in terms of distinct logical functions. The management and orchestration of the network-slicing-based cellular systems will be responsible for mapping physical network resources to the virtual ones in both the access and core networks and manage the physical network based on the virtual slices. Such a new network operating paradigm also requires demand forecasts for efficient network management and optimization.

© Springer International Publishing AG, part of Springer Nature 2018
X. Cheng et al., *Mobile Big Data*, Wireless Networks,
https://doi.org/10.1007/978-3-319-96116-3_6

For example, one of SON functions in virtualized cellular networks is to switch off a subset of cells when traffic loads across the network are extremely low and switch them on when the demand loads are heavy [8, 9]. The direct benefit of cell on/off switching is the operational cost reduction without any degradation of subscribers' quality of experience. However, cell towers may not be able to switch on to meet the traffic demands in a real-time manner. As a result, the cell on/off switching requires a robust short-term demand prediction to ensure subscribers' quality of experience. In addition, the virtualized core networks could reassign the virtual network resources [5] to support the cell tower on/off switching for efficient network management based on demand forecasts. Another example of predictive network management applications is the proactive drone deployment for cellular traffic offloading [10, 11]. When the traffic demands are beyond the cell capacity due to a large crowd inflow, the unmanned aerial vehicles (UAV) for cellular traffic offloading could mitigate such data traffic congestion problem. However, the overhead and capital expenditures of UAVs would limit the UAV number and the capability of the real-time UAV deployment. That is, UAVs need to travel a certain time before they can start serve. Hence, the proactive UAV deployment [11] could predictively send the UAVs to the data traffic congestion area, based on the crowd flow observation and demand forecasts.

6.1.2 Objective and Approaches

In this chapter, we study the mobile demand forecasting, the foundation of predictive mobile network management. In the literature, the mobile traffic/demand forecasting schemes have been studied for traffic apprehension and prediction via the Holt-Winter's exponential smoothing technique [12], information theory [13], and the seasonal ARIMA model [14]. However, all these demand forecasting models only consider the temporal aspect via various time series models without taking into account the spatial relevancy of cells. Models of mobile demand forecasting accounting for the spatial relevancy have been recently studied based on deep learning [15, 16]. In these models, the temporal aspect of demand time series is commonly studied via the recurrent neural networks (RNNs), while the spatial relevancy is captured by various grid-based spatial models (e.g., convolutional neural networks (CNNs)).

However, the main challenge of applying grid-based spatial models to per-cell demand forecasting is the irregular spatial distribution of cells in the real-world setting. Generally, the cell towers are distributed in a network covered area according to the population density. That is, the distance between two cell towers is about 500 m in the urban area, but can reach 2000 m in the rural area. Hence, grid-based models [15, 16] cannot directly apply. To utilize the grid-based models, one first needs to redivide the network covered area into a uniform square grid, and then study the aggregated demands of multiple cell towers residing in each lattice. Such spatial area redivision and demand aggregation will lead to the loss of the spatial granularity and will significantly limit the applications to future cellular network management that requires variable spatial granularity.

To this end, we study a flexible *graph-based spatial model* for the per-cell demand forecasting without any spatial resolution degradation and data aggregation. First, we realize that the spatiotemporal analysis of the per-cell demand time series via the semivariogram [17] reveals that the relevancy between the demands of two cells relies on the spatial distance of the two cells. That is, the dependency level of two cells would decrease when their spatial distance increases. Hence, we can build a dependency graph characterizing the relevancy of cells based on their spatial distances. In others words, the per-cell demands generated at each cell tower could be regarded as signals generated at the vertices of a graph. In addition, not only the recent demand history is applied to forecast the future demands, but also the periodic history (e.g., day(s) ahead demands) are considered in order to obtain an accurate demand predictor.

With the dependency graph formulation, the recently developed *graph convolutional network* (GCN) [18, 19] and the *gated recurrent unit* (GRU) neural networks [20] are employed to characterize the spatial aspect and the temporal aspect for demand forecasting, respectively. The GRU network is a gated version of *recurrent neural networks* (RNNs) in deep learning, which is well known for its good performance on sequence modeling. In GCNs, the graph convolution operation, originated from signal processing theory on graphs[21–23], is employed to replace the matrix multiplication in the feedforward neural networks. The power of graph convolution results from the ideas of parameter sharing and sparse interaction as in the traditional convolutional neural networks (CNNs) [24]. The sparse interaction in per-cell demand prediction means that the demand prediction of one cell is only related to itself and its immediate neighbors in the dependency graph. The parameter sharing assumes that the model parameters are shared across all cells of the network.

We first formulate the demand forecasting problem as a one-step ahead demand prediction problem. The demand forecasts after one step in the future are dynamically generated by the one-step ahead predictor. Three models, namely the spatial-only (GCNs), the temporal-only (GRU), and the spatiotemporal (GCGRU), are studied. The graph convolutional GRU (GCGRU) [25] is the model replacing the matrix multiplication operation with the graph convolution operation in GRU, inspired by the convolutional sequence modeling [26]. Compared with GCGRU, GRU without the embedded spatial information will predict the demand of one cell based on all other cells in the network, which would lead to an inferior generalization performance. Experiments show that the temporal-only model GRU could achieve a superior performance for the very-short-term demand forecasting for its much larger model capacity, but rapidly deteriorates when the forecast horizon increases. This results from the inferior generalization performance of GRU and the accumulated errors in the generated predicts. The GCGRU with the spatial and the temporal aspects modeled will generally have a superior forecast performance except for the very-short-term one.

6.2   Per-Cell Demand Time Series

6.2.1   Signaling Dataset

The signaling data is collected near the radio access networks (RAN) in a cellular network, which records communication events as well as location update events on all active subscribers in mobile networks. Data fields of the signaling data include (1) *subscriber's anonymized identifier*, (2) *time stamp* (e.g., 20160101184312), (3) *location coordinates* (i.e., the longitude and latitude of cell towers), (4) *event type*, and (5) *cell type* (i.e., small cell or macro cell). The longitude and latitude coordinates where the cell tower is located are accurate to six decimal places and time stamps are accurate to seconds. The signaling data logs event type as well as the direction of the event (e.g., initiating a call or being called). In the studied dataset, more than 6000 cells in total including small and macro cells with millions of subscribers are recorded in the studied dataset, as shown in Fig. 6.1. In the studied dataset, the average daily active subscribers is about three million. The time period of the studied signaling data is 104 days, from August 22, 2016 to December 3, 2016.

6.2.2   Per-Cell Demand Time Series

Based on the studied signaling dataset, two categories of service demands could be extracted, namely communication demands and tracking demands. The communication demands include the first four events on calls and texts recorded in the signaling dataset, to forecast which is the very task of this case study. The tracking demands

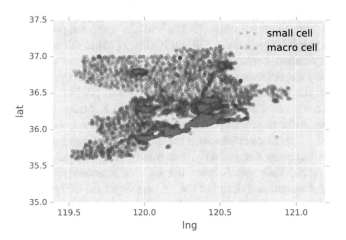

Fig. 6.1 Cell distribution map

could be obtained based on the location update events, which is closely related to crowd mobility. The location update frequency is once per hour, which may be too coarse to exactly describe the crowd flow, especially in the urban area (where cells are densely distributed). Hence, we focus on the communication demand forecasting in this case study.

With the spatiotemporal information of each event recorded, we define the per-cell demand as the number of communication events occurring within a cell during an event counting time window ΔT. Hence, a per-cell demand time series could be generated as follows,

$$\left[x_t^n, x_{t-1}^n, x_{t-2}^n, \cdots, x_{t-l+1}^n, \cdots \right], \tag{6.1}$$

where $x_t^n = \ln(1 + c_t^n)$ is defined as the per-cell demand within time window $[t - \Delta T, t)$, where c_t^n is the number of communication events of the n-th cell during the counting time period. Here, we utilize the commonly used logarithm function $\ln(1 + x)$ to convent the integer event number domain to the real number domain of demands. Here, we mainly study the demand forecasting in terms of the 10-min counting time windows, i.e., $\Delta T = 10$.

It can be clearly observed that small cells are densely deployed in the studied urban area (green areas as shown in Fig. 6.1). In a heterogeneous cellular networks, small cells are designed to assist their corresponding macro cell by offloading data traffic, whose coverage is also relatively much smaller than that of macro cells. As a result, the communication demands of small cells is sparse, which is not of interest in this case study. Hence, we aggregate the demand of small cells to its corresponding macro cell, which is determined by their spatially closest macro cell based on the location information (i.e., the longitude and latitude of cell towers). In other words, we study the per-cell aggregated demands within a spatial area covered by a macro cell.

In Fig. 6.2, the per-cell demands with different cell types are illustrated, namely business, entertainment, and residency. In each subfigure, three demand time series with different counting time window are plotted, $\Delta T = 5$ min, $\Delta T = 10$ min, and $\Delta T = 20$ min. One can easily observe that the large counting time window could significantly reduce the noise of the per-cell demand time series, as the larger counting time window acts like a smoothing filter applied on the one generated by the small counting time window. However, such noise reduction is at the cost of lowering the temporal resolution of demand time series as well as that of demand forecasts. In addition, it can be easily observed that per-cell demands are strongly periodic in terms of calendar days, regardless of cell types. Another periodic effect, that the demands during weekends are obviously less than those during weekdays, could be observed from the demand time series of the business-type cell (Fig. 6.2a). Such effects would inspire the feature engineering for demand forecasting, which will be discussed in detail later.

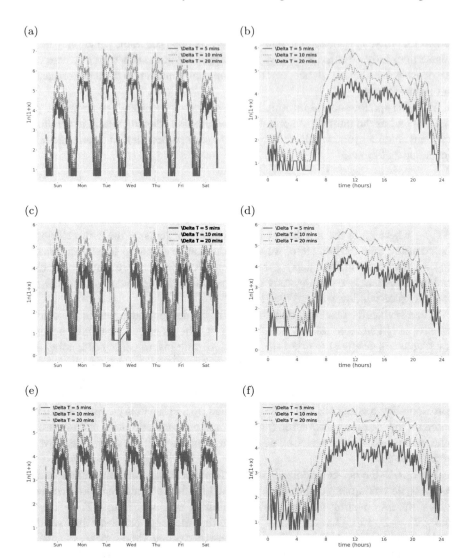

Fig. 6.2 Demand time series of various cell type, where the 7-day demands are recorded from November 27th, 2016 to December 3rd, 2016 and 24-h demands are recorded on November 27th, 2016. The business cell is located in the central business district (CBD), the entertainment-type cell is located in a public park, and the residence area is located in a large residential area. (**a**) 7-day demands, business, (**b**) 24-h demands, business, (**c**) 7-day demands, entertainment, (**d**) 24-h demands, entertainment, (**e**) 7-day demands, residency, (**f**) 24-h demands, residency

6.2.3 Analysis of Per-Cell Demand Time Series

Per-cell Demands Autocorrelation Analysis

We first investigate the autocorrelation analysis of the per-cell demands in a cell-wise manner, in order to determine the window length L should be taken into account for one-step ahead prediction (which will be discussed in detail latter). In the literature, the autocorrelation analysis and its partial derivative are commonly adopted to determine the order of auto regression integrated moving average (ARIMA) model. Specially, the autocorrelation function (ACF) would decide the order of the moving average, while the partial autocorrelation function (PACF) could shed lights on the order selection for the autoregression. Although the proposed time series model is quite different from ARIMA, the autocorrelation analysis could still be employed to suggest the window-length L selection.

Figure 6.3 shows the correlation analysis on the per-cell demand time series with the counting time window, $\Delta T = 10$. As the per-cell demand is strongly periodic with respect to calendar days as shown in Fig. 6.2, the per-cell demand of cell i can be further decomposed into two parts, mean and its random component,

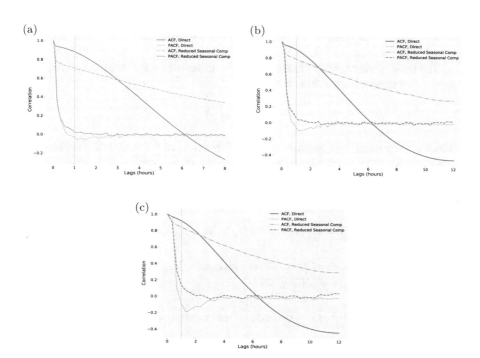

Fig. 6.3 Autocorrelation function and partial autocorrelation function of demand time series with event counting time windows $\Delta T = 5, 10, 20$ min (red solid curve: direct ACF; blue dot curve: direct PACF; purple dot-dash curve: ACF of reduced seasonal components; black dash curve: PACF of reduced seasonal components.). (**a**) $\Delta T = 5$ min, (**b**) $\Delta T = 10$ min, (**c**) $\Delta T = 20$ min

$$x_d^i = \bar{x}_d^i + \epsilon^i$$

where $\bar{x}_d^i$ is the periodic component. Hence, Fig. 6.3 shows two kinds of curves, namely the direct and the seasonal (periodic) component reduced, which demonstrate the (partial) autocorrelation analysis directly on the per-cell demand time series and the random component, respectively. It could be clearly observed that the PACF curves rapidly decrease to zero with the time lag increased, while the ACF curves are slowly decreasing, especially the direct one. One can conclude that 1 h history is sufficient for one-step ahead prediction, but longer history could benefit to capture the random component in the time series. As a result, we compare the different history lengths (half-hour, 1-h, 2-h, and 3-h) for all three proposed models in various settings.

Spatiotemporal Analysis

The objective of the spatiotemporal analysis on the multiple per-cell demand signals is to evaluate how demand signals vary in space and time. In other words, the correlation between two signals in terms of both the time lags and the spatial distance is of significant interest. Such spatiotemporal analysis would lead to our critical spatial modeling of demands observed by many cells irregularly spatially distributed.

The per-cell demand time series (6.1) could be expressed in terms of both the spatial and temporal aspects as follows,

$$x(s_n, t) = x_t^n \tag{6.2}$$

where s_n represents the detailed spatial information of the n-th cell (i.e., location coordinates). The semivariogram $\gamma(h)$ is a function employed to describe the spatial dependence of two stochastic processes generated in two locations s_n and s_m separated at h distance,

$$\gamma(h) = \mathrm{E}\left[(x(s_n) - x(s_m))^2 |\mathrm{dist}(s_n, s_m) = h\right] .$$

With the temporal dependence considered, the time lag τ should be further considered atop the spatial variogram $\gamma(h)$.

$$\gamma(h, \tau) = \mathrm{E}\left[(x(s, t) - x(s + h, t + \tau))^2\right]$$

However, the cell towers are distributed irregularly in the covered area according to the population density. Hence, we analyze the multiple per-cell demand processes in terms of the empirical spatiotemporal semivariogram [17, 27] as follows,

$$\gamma(h(l), \tau) = \frac{1}{|\mathcal{N}(h(l), \tau)|} \times \sum_{\substack{(n,m,t,t') \\ \in \mathcal{N}(h(l),\tau)}} \left[x(s_n, t) - x(s_m, t')\right]^2, \tag{6.3}$$

where

$$\mathcal{N}(h(l), \tau) = \{(n, m, t, t') | \text{dist}(s_n, s_m) \in h(l), |t - t'| = \tau\}.$$

The $\mathcal{N}(h(l), \tau)$ is a set to collect any signal pairs spatially separated at distance within the distance tolerance $h(l)$ and time lag τ. The distance tolerance $h(l)$ is employed to discretize the continuous spatial distance. Here, we utilize a linear uniform discretization with the spatial resolution 0.5 km. As a result, $h(l) = [(l - 1) \times 0.5, l \times 0.5)$.

Hence, the semivariogram, originated from spatial statistics, is employed to analyze the per-cell demands. Figure 6.4a, c show the semivariogram of per-cell demand time series in terms of both the spatial and temporal lags with different

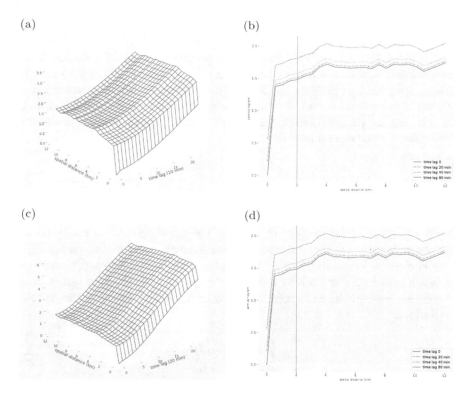

Fig. 6.4 Semivariogram of demand time series with different counting time windows. (**a**) $\Delta T = 10$ min, 3D, (**b**) $\Delta T = 10$ min, Project to 2D, (**c**) $\Delta T = 20$ min, 3D, (**d**) $\Delta T = 20$ min, Project to 2D

counting time window lengths $\Delta T = 10$ and $\Delta T = 20$, respectively. Based on the definition of semivariogram (6.3), the small value of semivariogram indicates the relatively high relevancy between signals separated at distance h and time lag τ. It could be observed that the semivariogram slowly grows along the time lag axis when $h = 0$, which suggests that the current per-cell demand is high correlated with its own history. As for the spatial dependence, it can be observed in both Fig. 6.4b, d that the value of semivariogram will stay the same after the spatial distance is 4 km. Such flatness suggests that any two cells with the distance larger than 4 km could be considered as irrelevant.

6.3   Demand Prediction Problem Formulation

With the definition of per-cell demands, the demand forecasting is aimed to predict the per-cell demands of all cells in a mobile network based on its history. Here, the demand forecasting is studied as the one-step ahead prediction problem as follows,

$$\hat{x}_{t+1} = f\left(x_t, x_{t-1}, \cdots, x_{t-l+1}, \cdots\right) \tag{6.4}$$

where $x_t = [x_t^1, x_t^2, \cdots, x_t^N]^T$ denotes the per-cell demands of cells across the covered area at time t and N is the total number of macro cells in the network. Hence, the prediction problem essentially amounts to the estimation of a function or predictor f based on the collected history data and the knowledge of cell locations. In this section, we will discuss the one-step ahead demand prediction with the innovative spatiotemporal modeling.

6.3.1   Graph-Based Spatial Relevancy Formulation

By the spatiotemporal analysis of multiple per-cell demand time series, it can be concluded that the demand relevancy between two cells declines when their spatial distance increases. Hence, we first propose to model the spatial relevancy between cells in the network by a dependency graph. The adjacency matrix A of the dependency graph can be obtained based on the spatial distance between cells as follows,

$$A_{ij} = \begin{cases} 1, & \text{dist}(s_i, s_j) \leq \zeta \\ 0, & \text{otherwise} \end{cases}, \tag{6.5}$$

where s_i denotes the location of cell i and ζ is the threshold, a hyperparameter that could be tuned. We set $\zeta = 2$ km in this case study. In fact, the threshold suggests that any two cells whose distance is beyond the threshold will be considered irrelevant. Such graph modeling could successfully make the cell relevancy sparse (from N^2 to $\sum_{i,j} A_{i,j}$), which can lead to a good demand forecasting generalization performance with the graph embedded in the predictor as detailed in Sects. 6.4 and 6.5. As a result, each cell could be regarded as a vertex in the spatial relevancy graph and the per-cell demand x_t is viewed as the signal observed at each vertex of the graph at time t.

6.3.2   Periodicity-Based Temporal Features

As shown in Fig. 6.2, it is obvious that the per-cell demand time series is periodic with respect to calendar days or weeks. In fact, such periodicity could provide valuable information for one-step ahead per-cell demand prediction at time t.

To predict x_{t+1}^i, not only the recent demand history $[x_t^i, x_{t-1}^i, \cdots, x_{t-L+1}^i]$ of cell i is taken into accounts, but also their corresponding days ahead demand observations will be regarded as input features for a predictor. Here, we only take the 1-day ahead and 6-day ahead observations as the extra features in order to make the predictor more dependent on the current trend. Hence, the input features of all cells in the network at time t take the form,

$$Z_t = [z_t^1, z_t^2, \cdots, z_t^N]^T, \tag{6.6}$$

where z_t^i denotes the input features of cell i at time t, i.e.,

$$z_t^i = [x_t^i, x_{t-n_d}^i, x_{t-6n_d}^i].$$

And n_d denotes the number of per-cell demand observations in one calendar day.

6.3.3   Graph-Sequence Demand Prediction Formulation

Based on the spatial and temporal modeling discussed above, the one-step ahead demand prediction problem could be further expressed as

$$\hat{x}_{t+1} = f(Z_t, Z_{t-1}, \cdots, Z_{t-L+1}; A) \tag{6.7}$$

where L is the length of recent history used for demand prediction. We will discuss the selection of L in Sect. 6.5. In this case study, we employ the commonly used mean absolute predicted error (MAE) as the evaluation criterion and cost function. Hence, the demand prediction problem could be expressed as follows,

$$\min_{f} \frac{1}{N} \sum_{n=1}^{N} |\hat{x}_{t+1}^{n} - x_{t+1}^{n}|. \tag{6.8}$$

Next, we will discuss the proposed per-cell demand predictor with effective graph and sequence information embedded based on deep learning.

6.4   Deep Graph-Sequence Spatiotemporal Modeling

The graph-based (GCN) model and the sequence-based model (GRU) are first proposed to individually capture the spatial and temporal aspects, respectively. In addition, we study their integrated version (GCGRU), which embeds the graph information in sequence models.

6.4.1   Spatial Modeling: Graph Convolutional Networks

Graph Filters and Graph Convolutions

The graph signal processing (GSP) [21–23] is recently developed to deal with signals generated from a graph, such as social networks and sensor networks, which is a general extension of the traditional signal processing techniques from regular sampled data (e.g., audio or image) to the irregular data (social network data). The graph signal processing combines both the signal processing and graph spectral theory, to fulfill the standard signal processing operations on the graph, e.g., convolution, filtering, translation, etc.

The main objective of building a spatial dependence graph in this case study is to predict the demand of one cell not only based on the its own demand history but also taking the demand history of its neighbors into account. In the graph signal processing theory, such objective could be captured by the graph Laplacian operation,

$$(\boldsymbol{L} \cdot \boldsymbol{x}_t)_i = \sum_{j \in \mathcal{N}_i} \left[x_t^i - x_t^j \right], \tag{6.9}$$

where $\boldsymbol{L} = \boldsymbol{D} - \boldsymbol{A}$ is the graph Laplacian and $\boldsymbol{D}$ is the diagonal matrix, i.e., $\boldsymbol{D}_{ii} = \sum_j A_{ij}$, recording the degree of each vertex in the graph. Intuitively, the graph Laplacian operation is essentially to capture the information of one vertex and its immediate neighbors.

Analogous to the filter design in the traditional signal processing, a graph filter could be expressed as polynomials in terms of the graph Laplacian [21],

$$g_\theta(\widetilde{L}) = \theta_0 I + \theta_1 \widetilde{L} + \theta_2 \widetilde{L}^2 + \cdots + \theta_K \widetilde{L}^K, \tag{6.10}$$

where $\widetilde{L}$ is the normalized graph Laplacian, i.e., $\widetilde{L} = I - D^{-\frac{1}{2}} A D^{-\frac{1}{2}}$. And θ_k is the filter coefficient of tap k. The order of graph filters would determine the order of neighbors of vertices in the graph affected by the filter.

By the eigendecomposition on the graph Laplacian, $\widetilde{L} = U \Lambda U^T$, any graph signal could be transformed to the corresponding graph spectral domain, $X = U x$, analogous to the discrete Fourier transform [21–23], where the eigenvectors U are viewed as a basis. As a result, the graph filter could be further expressed in the graph spectral domain,

$$g_\theta(\Lambda) = \theta_0 + \theta_1 \Lambda + \theta_2 \Lambda^2 + \cdots + \theta_K \Lambda^K. \tag{6.11}$$

Hence, the graph convolution operation $g_\theta(\widetilde{L}) * x_t$ can be calculated as multiplication operations in the graph spectral domain,

$$g_\theta(\widetilde{L}) \star x_t = U g_\theta(\Lambda) U^T x_t. \tag{6.12}$$

For details of graph signal processing, readers are recommended to refer to the survey paper [28].

Graph Convolutional Networks

The graph convolution would relate the signal of one vertex to others in terms of the graph topology, where the corresponding graph filter coefficients could be trainable based on data. As only the immediate neighbors are considered, the first-order graph filter based on (6.11) , $g_\theta^{(1)}(\Lambda) = \theta_0 + \theta_1 \Lambda$, is considered. In [18], Kipf and Welling proposed a simple first-order graph filer approximation based on Chebyshev polynomials of first kind [19] by forcing $\theta = \theta_0 = -\theta_1$ as follows,

$$g_\theta^{(1)}(\widetilde{L}) \star x_t = \widetilde{D}^{-\frac{1}{2}} \widetilde{A} \widetilde{D}^{-\frac{1}{2}} x_t \theta \tag{6.13}$$

where $\widetilde{A} = I + A$ and $\widetilde{D}$ is a diagonal matrix, $\widetilde{D}_{ii} = \sum_j A_{ij}$. Intuitively, the graph convolution operation (6.13) based on the first-order graph filter is essentially to first learn the shared filter coefficient θ by data in a vertex-wise manner and then to average the response of the vertex and its immediate neighbors based on the sparse connectivity information by the graph A.

Therefore, a graph convolutional network could be built based on the approximated first-order graph convolution operation to replace the matrix multiplication in the feedforward neural networks, which embeds the prior knowledge of graph topology into the learning model. As a result, each layer of graph convolutional networks is defined as

Spatial Model

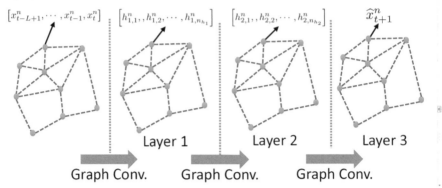

Fig. 6.5 Spatial modeling: graph convolutional networks (GCN)

$$H^{l+1} = \sigma\left(\widetilde{D}^{-\frac{1}{2}} \widetilde{A} \widetilde{D}^{-\frac{1}{2}} H^l \Theta^l\right) \tag{6.14}$$

where $\sigma(\cdot)$ denotes the activation function for nonlinearity modeling. $H^l \in \mathcal{R}^{N \times n_l}$ denotes the inputs of the l-th layer and $\Theta^l \in \mathcal{R}^{n_l \times n_{l+1}}$ is the trainable parameters in the model. Again, N denotes the number of vertices of the graph. In each graph convolution operation, the $H^l \Theta^l$ in (6.14) is first to learn the pattern in a cell-wise manner with shared parameters Θ^l. The product of $H^l \Theta^l$ and $\widetilde{D}^{-\frac{1}{2}} \widetilde{A} \widetilde{D}^{-\frac{1}{2}}$ is essentially equivalent to the weighted sum over the cell and its first-order neighbors.

In the context of the per-cell demand prediction problem, we propose a three-layer graph convolutional network as the demand predictor f as detailed in Model 1 and Fig. 6.5.

Model 1 (Graph Convolutional Networks (GCN)) *A per-cell demand predictor is approximated by a three-layer graph convolutional network,* $\hat{x}_{t+1} = \hat{f}(\mathbf{Z}_t^{(GCN)}, A)$, *i.e.,*[1]

$$Layer\ 1: \quad H^{(1)} = \sigma\left(\widehat{A} \mathbf{Z}_t^{(GCN)} \Theta^{(1)}\right), \quad \Theta^{(1)} \in \mathcal{R}^{(L \times F) \times n_1}$$

$$Layer\ 2: \quad H^{(2)} = \sigma\left(\widehat{A} H^{(1)} \Theta^{(2)}\right), \quad \Theta^{(2)} \in \mathcal{R}^{n_1 \times n_2} \tag{6.15}$$

$$Layer\ 3: \quad \hat{x}_{t+1} = \widehat{A} H^{(2)} \Theta^{(3)}, \quad \Theta^{(3)} \in \mathcal{R}^{n_2 \times 1}$$

where $\widehat{A} = \widetilde{D}^{-\frac{1}{2}} \widetilde{A} \widetilde{D}^{-\frac{1}{2}}$ *and* $\mathbf{Z}_t^{(GCN)}$ *denotes the input of the GCN with L-length window,*

---

[1]For simplicity, we omit the bias of all models.

Here, $\mathbf{Z}_t^{(GCN)}$ is the L-length demand history with days ahead features as the input, i.e.,

$$\mathbf{Z}_t^{(GCN)} = [\mathbf{Z}_t, \cdots, \mathbf{Z}_{t-L+1}].$$

In other words, the L-length demand history and extra days ahead features of each cell are regarded as its input features of GCNs without explicit sequence modeling. As a result, the total number of free trainable parameters in the proposed three-layer GCN is $n_{h_1}(L \times F) + n_{h_1}n_{h_2} + n_{h_2}$. Here, two-layer graph convolution operations are employed in graph-based models to mimic second order graph filter based on the simple first-order graph filter approximation. Accordingly, we set the threshold ζ in (6.5) to be 2 km to capture the neighbors within 4 km after two-layer graph convolution operations.

6.4.2   Temporal Modeling: Gated Recurrent Unit (GRU) Networks

In the literature, the recurrent neural networks (RNNs) have been proved to be an effective sequence model [29], which is designed to capture the sequential information inherited in data, e.g., audio, nature language, etc. Essentially, RNNs add a feedback path in the feedforward neural networks, which could provide the information of the previous inputs so that the current output not only depends on the current inputs but also relies on the hidden state learned from previous inputs as follows,

$$\boldsymbol{h}_t = \sigma\left(\boldsymbol{W}\boldsymbol{z}_t + \boldsymbol{V}\boldsymbol{h}_{t-1}\right), \tag{6.16}$$

where $\boldsymbol{h}_{t-1}$ denotes the hidden states updated previously. The basic issue of RNNs is the vanishing or exploding gradients after the gradient propagating many stages during training, which would largely degrade the performance of RNN. In RNNs, the hidden state is employed to remember the current state of the cell as well as the cell output, which could be further fed to next layer of the network. The gated recurrent unit (GRU) network is one of the specially designed RNNs aimed to solve the gradient vanishing problem, which has a capability of controlling the updating process by adding two gates, namely the reset gate $\boldsymbol{g}_r$ and the update gate $\boldsymbol{g}_u$ in a GRU cell; that is,

$$\begin{aligned} \boldsymbol{r}_t &= \sigma\left(\boldsymbol{W}_r\boldsymbol{z}_t + \boldsymbol{V}_r\boldsymbol{h}_{t-1}\right) \\ \boldsymbol{u}_t &= \sigma\left(\boldsymbol{W}_u\boldsymbol{z}_t + \boldsymbol{V}_u\boldsymbol{h}_{t-1}\right) \end{aligned}, \tag{6.17}$$

where $\sigma(\cdot)$ denotes the sigmoid function. These two gates control how much information should be passed through in different places of GRU cells as follows,

$$h'_t = \tanh\left(W_h z_t + r_t \circ V_h h_{t-1}\right)$$

$$h_t = u_t \circ h_{t-1} + (1 - u_t) \circ h'_t \qquad \text{(6.18)}$$

where c_t and h_t denote the cell state and the hidden state at time t, respectively. Here, the operator "$\circ$" denotes the element-wise multiplication. As suggested in (6.18), the reset gate is employed to decide how much pass information should be forgotten in the current state update, while the update gate is utilized to determine how much the past information and the current input information should be drawn to generate the final hidden state. Details of the GRU cell structure are shown in Fig. 6.6.

Hence we propose a three-layer many-to-one GRU network as a per-cell demand predictor as described in Model 2, which regards the per-cell demand of all cells at each time stamp as inputs.

Model 2 (Gated Recurrent Unit (GRU)) *A per-cell demand predictor is approximated by a three-layer GRU network with two GRU layers and one full-connection layer. The GRU sequence model is demonstrated in Fig. 6.7 and illustrated mathematically as follows,*

$$\textit{Layer 1: } \; h_t^{(1)} = \eta_{gru}^{(1)}\left(z_t^{(GRU)}, h_{t-1}^{(1)}\right)$$

$$\textit{Layer 2: } \; h_t^{(2)} = \eta_{gru}^{(2)}\left(h_t^{(1)}, h_{t-1}^{(2)}\right) \qquad \text{(6.19)}$$

$$\textit{Layer 3: } \; \hat{x}_{t+1} = W^{(3)} h_t^{(2)}$$

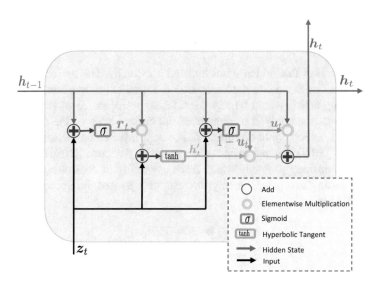

Fig. 6.6 Details of GRU cell structure

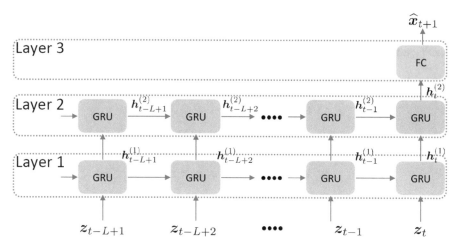

Fig. 6.7 Temporal modeling: gated recurrent units (GRU)

where $\eta_{gru}^{(i)}(\cdot, \cdot)$ denotes the function of the layer i GRU cell as described in (6.17) and (6.18), in which the trainable parameters are listed as follows,

$$\text{Layer 1: } \boldsymbol{W}_{r,u,h}^{(1)} \in \mathcal{R}^{(N \times F) \times n_{h_1}}, \boldsymbol{V}_{r,u,h}^{(1)} \in \mathcal{R}^{n_{h_1} \times n_{h_1}}$$

$$\text{Layer 2: } \boldsymbol{W}_{r,u,h}^{(2)} \in \mathcal{R}^{n_{h_1} \times n_{h_2}}, \boldsymbol{V}_{r,u,h}^{(2)} \in \mathcal{R}^{n_{h_2} \times n_{h_2}}$$

$$\text{Layer 3: } \boldsymbol{W}^{(3)} \in \mathcal{R}^{n_{h_1} \times n_{h_2}},$$

where n_{h_1} and n_{h_2} denote the size of hidden states in layer 1 and layer 2, respectively.

Here, the input $z_t^{(GRU)}$ is a vector that contains features of all cells at time t, whose size is $(N \times F) \times 1$. As a result, the number of trainable parameters in Model 2 is $3n_{h_1}(N \times F + n_{h_1}) + 3n_{h_2}(n_{h_1} + n_{h_2}) + n_{h_2}N$. In GRU, we only model the temporal aspect of the per-cell demand data, but omit the spatial information. In other words, the spatial local dependence is not considered in the GRU model, but the full connection from one cell to all other cells are taken into account, which may lead to overfitting in the GRU model.

6.4.3   Spatiotemporal Modeling: Graph Convolutional GRU (GCGRU)

With the spatial and temporal information modeled, the GRU and GCN can be integrated to utilize both the spatial and temporal information, which is termed as graph convolutional GRU (GCGRU). In GCGRU, the global connection among vertices (matrix multiplication in GRUs) is replaced by the local graph convolution (6.14) in each gate as follows,

$$\begin{aligned} \boldsymbol{R}_t &= \sigma\left(\widehat{\boldsymbol{A}}(\boldsymbol{Z}_t\boldsymbol{\Theta}_r + \boldsymbol{H}_{t-1}\boldsymbol{\Psi}_r)\right) \\ \boldsymbol{U}_t &= \sigma\left(\widehat{\boldsymbol{A}}(\boldsymbol{Z}_t\boldsymbol{\Theta}_u + \boldsymbol{H}_{t-1}\boldsymbol{\Psi}_u)\right) \end{aligned} \tag{6.20}$$

where $\boldsymbol{G}_{r,u} \in \mathcal{R}^{N \times n_h}$. Also, the hidden states are also updated locally as follows,

$$\begin{aligned} \boldsymbol{H}'_t &= \tanh\left(\widehat{\boldsymbol{A}}(\boldsymbol{Z}_t\boldsymbol{\Theta}_c + \boldsymbol{R}_t \circ \boldsymbol{H}_{t-1}\boldsymbol{\Psi}_c)\right) \\ \boldsymbol{H}_t &= \boldsymbol{U}_t \circ \boldsymbol{H}_{t-1} + (1 - \boldsymbol{U}_t) \circ \boldsymbol{H}'_t \end{aligned} \tag{6.21}$$

Accordingly, a per-cell demand predictor based on GCGRU cell is proposed to model both the spatial and temporal dimension of the per-cell demand time series as illustrated in Model 3 (Fig. 6.8).

Model 3 (Graph Convolutional GRU (GCGRU)) *A per-cell demand predictor is approximated by a three-layer GCGRU with two layers of GCGRU cells and one graph convolutional layer, i.e.,*

$$\textit{Layer 1: } \boldsymbol{H}_t^{(1)} = \eta_{gcgru}^{(1)}\left(\boldsymbol{Z}_t, \boldsymbol{H}_{t-1}^{(1)}\right)$$

$$\textit{Layer 2: } \boldsymbol{H}_t^{(2)} = \eta_{gcgru}^{(2)}\left(\boldsymbol{H}_t^{(1)}, \boldsymbol{H}_{t-1}^{(2)}\right) \tag{6.22}$$

$$\textit{Layer 3: } \hat{\boldsymbol{x}}_{t+1} = \widehat{\boldsymbol{A}}\boldsymbol{H}_t^{(2)}\boldsymbol{\Theta}^{(3)}$$

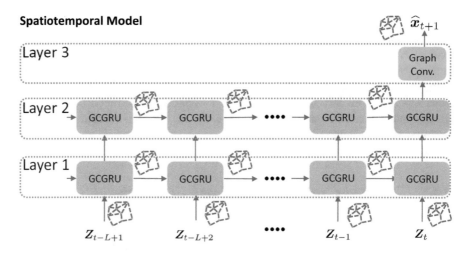

Fig. 6.8 Spatiotemporal modeling: graph convolutional GRU (GCGRU)

	Model Type	Input Dim.	Feature Size	Param. Num.	Example
GCN	Spatial	Vertex Based 2D Matrix	L x F	$n_{h_1} \times (L \times F)$ $+ n_{h_2} \times n_{h_1}$ $+ n_{h_2}$	2,208
GRU	Temporal	Time Based 1-D Vector	N x F	$3n_{h_1} \times (N \times F + n_{h_1})$ $+ 3n_{h_2} \times (n_{h_1} + n_{h_2})$ $+ n_{h_2} \times N$	238,976
GCGRU	Spatio-temporal	Time Based 2-D Matrix	F	$3n_{h_1} \times (n_{h_1} + F)$ $+ 3n_{h_2} \times (n_{h_2} + n_{h_1})$ $+ n_{h_2}$	9,536

L: length of sequence (12) n_{h1} Number of hidden units in the 1st layer (32)
F: features of each cell at each time (3) n_{h2} Number of hidden units in the 2nd layer (32)
N: number of cells (718)

Fig. 6.9 Summary of studied models

where $\eta_{gcgru}^{(i)}(\cdot, \cdot, \cdot)$ denotes the layer i GCGRU cell based on (6.20) and (6.21),where the trainable parameters are illustrated as follows,

$$Layer\ 1:\ \boldsymbol{\Theta}_{r,u,h}^{(1)} \in \mathcal{R}^{F \times n_{h_1}}, \boldsymbol{\Psi}_{r,u,h}^{(1)} \in \mathcal{R}^{n_{h_1} \times n_{h_1}}$$

$$Layer\ 2:\ \boldsymbol{\Theta}_{r,u,h}^{(2)} \in \mathcal{R}^{n_{h_1} \times n_{h_2}}, \boldsymbol{\Psi}_{r,u,h}^{(2)} \in \mathcal{R}^{n_{h_2} \times n_{h_2}}.$$

$$Layer\ 3:\ \boldsymbol{\Theta}^{(3)} \in \mathcal{R}^{n_{h_2} \times 1}$$

Again, n_{h_1} and n_{h_2} denote the size of hidden states in layer 1 and layer 2, respectively.

Here, the input $\boldsymbol{Z}_t$ at time t is a matrix with the shape $N \times F$ defined by (6.6). The number of trainable parameters is $3n_{h_1}(n_{h_1} + F) + 3n_{h_2}(n_{h_2} + n_{h_1}) + n_{h_2}$. Compared with GRU, the number of trainable parameters could be largely reduced, since the parameters are shared across the graph with local dependence modeled. Such parameter sharing could mitigate the overfitting problem by structurally shrinking the capacity of the model. Details of model comparisons are summarized in Fig. 6.9.

6.5   Experiments

In this section, we verify three proposed spatial, temporal and spatiotemporal models based on the extracted per-cell demand data of 718 cell towers in the mobile network. The per-cell demands are first normalized by their mean and standard deviation in a cell-wise manner. The demand predictors proposed are implemented by PyTorch [30], which is a deep learning framework with automatic differentiation and dynamic computational graph. The training dataset is from August 22, 2016 to November 26, 2016 and the test dataset is from November 27, 2016 to December 3, 2016.

In this case study, we employ the mean absolute error (MAE) as the criterion to evaluate the predictors studied. Though the forecast problem is formulated as a one-step ahead prediction problem (6.4), the per-cell demand predictor should be capable of forecasting the demands of a future time window. In fact, the demand forecasting is fulfilled by the dynamic prediction via the one-step ahead predictor, which would take predicted demands as inputs to further forecast the future demands, e.g., $\hat{x}_{t+2} = f(\hat{x}_{t+1}, x_t, \cdots)$.

As a results, two parameters, forecast horizon and forecast resolution, are important for a forecasting problem. The forecast resolution relies on the length of event counting time window, which is a predict per 10 min. We focus on the studied models with the forecasting horizon of 24 h. In [14], a seasonal ARIMA model is proposed to predict the per-cell demands of a single cell with seasonal component modeled, SARIMA $(1, 0, 3) \times (1, 1, 1)$,

$$
\begin{aligned}
(1 - ar_1 z^{-1})&(1 - sar_1 z^{-n_d})(1 - z^{-n_d}) x_t^i \\
&= (1 + ma_1 z^{-1} + ma_2 z^{-2} + ma_3 z^{-3})(1 + sma_1 z^{-n_d}) \epsilon_t,
\end{aligned}
$$

where ϵ_t denotes the noise component and z_{-1} denotes the operation of one time lag. Though SARIMA cannot model the spatial correlation among cells nor simultaneously predict the per-cell demands across the entire network, we could still perform the comparisons in a cell-wise manner. In Fig 6.10, an example of 24-h demand forecasting of a cell is showed, including the proposed models and the SARIMA. It could be clearly observed that the predicts by the SARIMA is more fluctuate than that of our models, while our proposed models smoothly trace the ground truth curve. In Fig. 6.11, the average predicted MAE comparisons among three proposed models over all cells in the network is demonstrated. Overall, the spatiotemporal model (GCGRU) is the best except for the case that forecast horizons are less than 5 h. As the capacity of the GRU model without parameter sharing and locality modeling is much larger than the one of GCGRU, demonstrated by their number of trainable parameters (see Fig. 6.9), the GRU can well capture the insight for one-step ahead prediction. However, the GRU also easily models the noise into the predictor during training, which could lead to the overfitting issue and worsen the forecasting performance of the model. Figure 6.13 also demonstrates the our proposed GCGRU model performs better than the SARIMA.

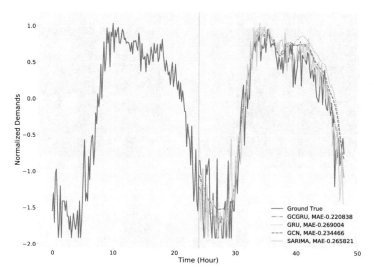

Fig. 6.10 A example of dynamic per-cell demand forecasting

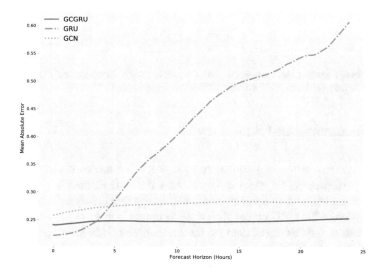

Fig. 6.11 MAE performance of dynamic forecasting over all cells

Figure 6.12 illustrates the differences of demand history length for per-cell demand prediction. Overall, the longer demand history could definitely improve the accuracy for large forecast horizon, especially for the GRU-based models, which may result from the hidden states of GRU-based models could remember more information when their hidden states are updated longer. On the other hand, the GCN model is not sensitive to the demand history length when $L \geq 6$ (longer than or equal to 1 h) due to the lack of explicit temporal modeling, as shown in Fig. 6.12a.

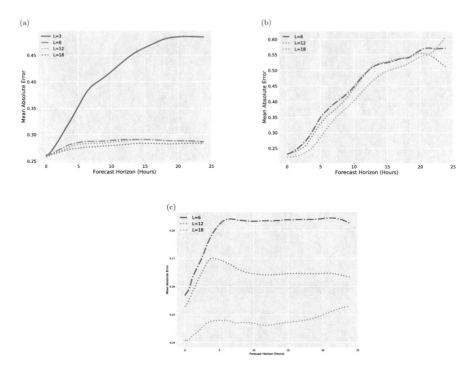

Fig. 6.12 MAE comparison between different window length L, where the event count time window is 10 min. (**a**) GCN, (**b**) GRU, (**c**) GCGRU

6.6   Discussions and Summary

As demonstrated in the experiment results, the GRU model could always have the best performance for the very-short-term demand forecasting, namely less than 3 h. However, due to accumulated error during dynamic prediction and week generalization of the GRU model, the GCGRU model is more capable for the short-term, mid-term and day ahead demand forecasting. The GCN is also stable for such forecast horizons but is less accurate, while the number of trainable parameters is much smaller as illustrated in Fig. 6.9. The SARIMA model (Fig. 6.13) performs well for the per-cell demands prediction task, but it is modeled in a cell-wise manner. That is, the per-cell demand needs to be predicted cell-by-cell. As a result, the parameters of SARIMA is linearly scaling with the number of cells in the network, while our proposed GCGRU takes both the spatial and temporal into accounts with fixed number trainable parameters and could have a relative small trainable parameters for a large mobile network.

To sum up, we study the per-cell demand forecasting in cellular networks. To deal with the irregular cell spatial distribution for spatial relevancy modeling among cells, we proposed to model the spatial relevancy among cells as a dependency graph based on spatial distances among cells without losing spatial granularity.

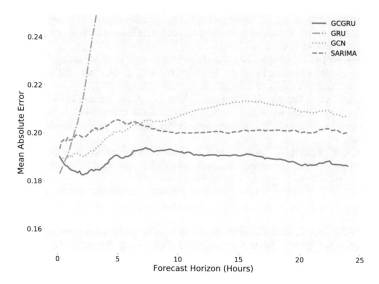

Fig. 6.13 Compared with SARIMA $(1, 0, 1) \times (1, 1, 1)$

Accordingly, we studied three models for demand forecasting, the spatial only (graph), the temporal only (sequence), and the spatiotemporal model (graph-sequence) based on deep learning. The spatiotemporal model simultaneously could capture both the spatial and temporal aspects in demand forecasting, which could achieve a superior forecasting performance demonstrated by experiment results.

References

1. F. Malandrino, C. F. Chiasserini, and S. Kirkpatrick, "Understanding the present and future of cellular networks through crowdsourced traces," in *Proceedings of the 18th International Symposium on A World of Wireless, Mobile and Multimedia Networks (WoWMoM)*, Macau, China, Jun. 12–15, 2017.
2. M. Peng, D. Liang, Y. Wei, J. Li, and H. H. Chen, "Self-configuration and self-optimization in LTE-advanced heterogeneous networks," *IEEE Communications Magazine*, vol. 51, no. 5, pp. 36–45, May 2013.
3. O. G. Aliu, A. Imran, M. A. Imran, and B. Evans, "A survey of self organisation in future cellular networks," *IEEE Communications Surveys Tutorials*, vol. 15, no. 1, pp. 336–361, First 2013.
4. P. V. Klaine, M. A. Imran, O. Onireti, and R. D. Souza, "A survey of machine learning techniques applied to self-organizing cellular networks," *IEEE Communications Surveys Tutorials*, vol. 19, no. 4, pp. 2392–2431, Fourthquarter 2017.
5. R. Li, Z. Zhao, X. Zhou, G. Ding, Y. Chen, Z. Wang, and H. Zhang, "Intelligent 5g: When cellular networks meet artificial intelligence," *IEEE Wireless Communications*, vol. 24, no. 5, pp. 175–183, Oct. 2017.

6. E. J. Kitindi, S. Fu, Y. Jia, A. Kabir, and Y. Wang, "Wireless network virtualization with SDN and C-RAN for 5G networks: Requirements, opportunities, and challenges," *IEEE Access*, vol. 5, pp. 19 099–19 115, Sep. 2017.

7. H. Zhang, N. Liu, X. Chu, K. Long, A. H. Aghvami, and V. C. M. Leung, "Network slicing based 5g and future mobile networks: Mobility, resource management, and challenges," *IEEE Communications Magazine*, vol. 55, no. 8, pp. 138–145, Aug. 2017.

8. H. Ghazzai, M. J. Farooq, A. Alsharoa, E. Yaacoub, A. Kadri, and M. S. Alouini, "Green networking in cellular hetnets: A unified radio resource management framework with base station on/off switching," *IEEE Transactions on Vehicular Technology*, vol. 66, no. 7, pp. 5879–5893, Jul. 2017.

9. M. Ismail, W. Zhuang, E. Serpedin, and K. Qaraqe, "A survey on green mobile networking: From the perspectives of network operators and mobile users," *IEEE Communications Surveys Tutorials*, vol. 17, no. 3, pp. 1535–1556, thirdquarter 2015.

10. Y. Zeng, R. Zhang, and T. J. Lim, "Wireless communications with unmanned aerial vehicles: opportunities and challenges," *IEEE Communications Magazine*, vol. 54, no. 5, pp. 36–42, May 2016.

11. P. Yang, X. Cao, C. Yin, Z. Xiao, X. Xi, and D. Wu, "Proactive drone-cell deployment: Overload relief for a cellular network under flash crowd traffic," *IEEE Transactions on Intelligent Transportation Systems*, vol. 18, no. 10, pp. 2877–2892, Oct. 2017.

12. D. Tikunov and T. Nishimura, "Traffic prediction for mobile network using Holt-Winter's exponential smoothing," in *Proceedings of the 15th International Conference on Software, Telecommunications and Computer Networks*, Split-Dubrovnik, Croatia, Sep. 2007.

13. R. Li, Z. Zhao, X. Zhou, J. Palicot, and H. Zhang, "The prediction analysis of cellular radio access network traffic: From entropy theory to networking practice," *IEEE Communications Magazine*, vol. 52, no. 6, pp. 234–240, Jun. 2014.

14. F. Xu, Y. Lin, J. Huang, D. Wu, H. Shi, J. Song, and Y. Li, "Big data driven mobile traffic understanding and forecasting: A time series approach," *IEEE Transactions on Services Computing*, vol. 9, no. 5, Sep. 2016.

15. J. Zhang, Y. Zheng, D. Qi, R. Li, and X. Yi, "Dnn-based prediction model for spatio-temporal data," in *Proceedings of the 24th ACM SIGSPATIAL International Conference on Advances in Geographic Information Systems*, Burlingame, California, Oct. 31 - Nov. 3, 2016, pp. 92:1–92:4.

16. J. Wang, J. Tang, Z. Xu, Y. Wang, G. Xue, X. Zhang, and D. Yang, "Spatiotemporal modeling and prediction in cellular networks: A big data enabled deep learning approach," in *Proceedings of IEEE Conference on Computer Communications (INFOCOM)*, Atlanta, GA, USA, May 1–4, 2017.

17. N. Cressie and H.-C. Huang, "Classes of nonseparable, spatio-temporal stationary covariance functions," *Journal of the American Statistical Association*, vol. 94, no. 448, pp. 1330–1339, 1999.

18. T. N. Kipf and M. Welling, "Semi-supervised classification with graph convolutional networks," Paris, France, Apr. 24–26, 2017.

19. M. Defferrard, X. Bresson, and P. Vandergheynst, "Convolutional neural networks on graphs with fast localized spectral filtering," in *Advances in Neural Information Processing Systems 29*, D. D. Lee, M. Sugiyama, U. V. Luxburg, I. Guyon, and R. Garnett, Eds., 2016, pp. 3844–3852.

20. J. Chung, C. Gulcehre, K. Cho, and Y. Bengio, "Empirical evaluation of gated recurrent neural networks on sequence modeling," *arXiv preprint arXiv:1412.3555*, 2014.

21. A. Sandryhaila and J. M. F. Moura, "Discrete signal processing on graphs," *IEEE Transactions on Signal Processing*, vol. 61, no. 7, pp. 1644–1656, Apr. 2013.

22. D. I. Shuman, S. K. Narang, P. Frossard, A. Ortega, and P. Vandergheynst, "The emerging field of signal processing on graphs: Extending high-dimensional data analysis to networks and other irregular domains," *IEEE Signal Processing Magazine*, vol. 30, no. 3, pp. 83–98, May 2013.

23. A. Sandryhaila and J. M. F. Moura, "Big data analysis with signal processing on graphs: Representation and processing of massive data sets with irregular structure," *IEEE Signal Processing Magazine*, vol. 31, no. 5, pp. 80–90, Sep. 2014.

24. I. Goodfellow, Y. Bengio, and A. Courville, *Deep Learning*. MIT Press, 2016.

25. Y. Seo, M. Defferrard, P. Vandergheynst, and X. Bresson, "Structured sequence modeling with graph convolutional recurrent networks," *eprint arXiv:1612.07659*.

26. X. Shi, Z. Chen, H. Wang, D.-Y. Yeung, W.-k. Wong, and W.-c. WOO, "Convolutional LSTM Network: A machine learning approach for precipitation nowcasting," in *Advances in Neural Information Processing Systems 28*, C. Cortes, N. D. Lawrence, D. D. Lee, M. Sugiyama, and R. Garnett, Eds., 2015, pp. 802–810.

27. X. Jian, R. A. Olea, and Y.-S. Yu, "Semivariogram modeling by weighted least squares," *Computers & Geosciences*, vol. 22, no. 4, pp. 387–397, May 1996.

28. A. Ortega, P. Frossard, J. Kovačević, J. M. Moura, and P. Vandergheynst, "Graph signal processing," *arXiv preprint arXiv:1712.00468*, 2017.

29. Y. LeCun, Y. Bengio, and G. Hinton, "Deep learning," *Nature*, vol. 521, no. 7553, pp. 436–444, May 2015.

30. A. Paszke, S. Gross, S. Chintala, G. Chanan, E. Yang, Z. DeVito, Z. Lin, A. Desmaison, L. Antiga, and A. Lerer, "Automatic differentiation in PyTorch," Long Beach, CA, USA, Dec. 9, 2017.

Chapter 7
Case Study: User Identification for Mobile Privacy

7.1 Background

7.1.1 Privacy Attack: User Identification

To facilitate the novel mobile data-driven applications and services as discussed in previous chapters, mobile big data with spatiotemporal information may need to be released to third parties or even to the public. However, direct data publishing may lead to a significant subscriber's privacy leakage risk [3], immediately resulting in data availability issues. To protect subscribers' privacy, the common practice is to anonymize the dataset by replacing subscribers' identifiers (e.g., name, social security number, etc.) with randomly generated strings. Moreover, the anonymized identifiers could be replaced frequently (e.g., every other month) in data management for further privacy protection. However, these practices may not be able to effectively protect subscriber's privacy due to the uniqueness of human mobility trajectory [2, 4–8]. Such uniqueness of subscribers' spatiotemporal trajectories does not directly lead to a privacy leakage, but does result in a high privacy leakage risk [9], which makes anonymization less effective.

Another privacy risk is also significantly concerned for mobile privacy protection. Specifically, the user identity may be revealed across two datasets based on subscribers' mobility patterns [2, 10]. The user identity linkage across two datasets will enrich the adversary's knowledge if the privacy attacker has already learned the user identity information of one of the datasets. In this work, we study the location privacy in terms of user identity linkage based on their spatiotemporal behaviors from the perspective of a privacy attacker. The main purpose of this work is to evaluate subscriber's privacy leakage risk in terms of user identifiability across two datasets via diverse semantic spatiotemporal feature extraction.

© Springer International Publishing AG, part of Springer Nature 2018
X. Cheng et al., *Mobile Big Data*, Wireless Networks,
https://doi.org/10.1007/978-3-319-96116-3_7

7.1.2 Approach: Multi-Feature Ensemble Matching Framework

In this case study, a *multi-feature ensemble matching framework* is studied to solve the user identification problem based on a k-cardinality minimum-cost bipartite matching (k-MinBM) [2]. The k-MinBM formulation roots in the exclusiveness assumption that no more than one record in a dataset is generated by one user. The exclusiveness assumption has been proved to be superior in the context of user identification problems [1]. In fact, the success of k-MinBM relies on an effective quantitative distance measure between two spatiotemporal attributes. Based on the formulation by k-MinBM, the studied multi-feature ensemble matching framework is aimed to answer two crucial questions as follows:

- How to model and extract various features that could distinctly represent a user and be employed to measure the distance between two spatiotemporal attributes from diverse perspectives;
- How to effectively integrate identification results based on multiple diverse spatiotemporal features to reduce false matched pairs and ensure a higher user identification accuracy, especially when the coexisting user number in two datasets is unknown.

Therefore, we first semantically explore and exploit the spatiotemporal features and their corresponding distance measures for user identification. In [2], the visiting frequency captured by the empirical probability distribution over the location point set is employed to distinctly represent a user, based on which the distance between two users is measured by Jensen-Shannon divergence on two empirical probability distributions. However, such data modeling completely discards the temporal information from the data and omits the geospatial relationship between location points. Hence, we jointly model the visiting frequency and location-dependent duration, in which we utilize the temporal information provided by the spatiotemporal trace. Each observed duration at a specific location point is assumed to be generated from an exponential distribution conditioned on the location point, where the empirical visiting probability acts as a prior. Furthermore, we explore a heuristic yet semantic representation, daily habitat regions, characterizing the maximum area that a user cover during a day. The daily habitat region feature is extracted by the convex hull of the whole visiting location point set within a day and the corresponding distance measure is proposed based on the cosine distance defined on the overlapping area of these two convex hulls.

With diverse spatiotemporal features exploited for user identification, we study an *ensemble matching* procedure to integrate results generated by diverse distance measures, which could largely reduce false matching. In fact, the integration of diverse spatiotemporal features characterizing a user in different aspects is anticipated to achieve a better performance. However, classical ensemble learning approaches in the literature such as boosting and bagging may not be applied, for the bipartite matching in user identification is a different problem. In fact, our proposed ensemble matching approach acts as a information/result fusion inspired by the "stacking" approach [11]. The proposed ensemble matching is to first generate matching candidates via k-MinBM on each distance measure independently, based on which an ensemble procedure is proposed to filter out the final matching with majority votes and the exclusiveness assumption ensured. Experiments confirm that our proposed ensemble matching could achieve superior performance, compared with those without ensemble, especially when the coexisting user number k is small.

7.2   Problem Description

Assume that a spatiotemporal dataset $\mathcal{X}$ is collected by a mobile network operator during a time period, in which each subscriber is represented as a tuple $(i, X_i) \in \mathcal{X}$. The i denotes the i-th subscriber in $\mathcal{X}$ pinpointed by a distinct anonymized identifier, while X_i represents his/her corresponding collection of spatiotemporal records. The spatiotemporal attribute X_i is a timestamped location point sequence, recording when and where the subscriber interacts with the mobile network,

$$X_i = [(t_1, x_1), \cdots, (t_h, x_h), \cdots], \ x_l \in \mathcal{A}, \tag{7.1}$$

where t_h denotes the timestamp of the h-th event. And x_h is the location where the h-th event occurs, drawn from a location point set $\mathcal{A}$. The location point set $\mathcal{A}$ is a collection of base stations spatially distributed in an area covered by the mobile network, where its letter $a_l \in \mathcal{A}$ can represent the identifier of the base station l or its GPS coordinate, i.e., $a_l = (\text{lng}_l, \text{lat}_l)$. A typical example of such data is the commonly studied call detail records (CDR) [3], which are call or texting event logs collected by network operators for service charging.

Assume that the privacy attacker can access two of such spatiotemporal datasets, $\mathcal{X}$ and $\mathcal{Y}$, collected in two non-overlapping time periods. The privacy attacker has already obtained user identity information from dataset $\mathcal{Y}$, then he/she attempts to connect the spatiotemporal information generated by the same user from the datasets $\mathcal{X}$ and $\mathcal{Y}$, despite that these spatiotemporal attributes are associated with different anonymized identifiers in these two datasets. The studied user identification problem in this work is henceforth defined as connecting spatiotemporal attributes across two datasets (with different identifiers) corresponding to the same users so that attacker's knowledge on users of interests could be enriched.

7.2.1   Assumptions

Two assumptions are generally made for the studied user identification problem.

- **Consistency.** An individual user's spatiotemporal behavior is assumed to be statistically consistent to some degree in datasets $\mathcal{X}$ and $\mathcal{Y}$. In other words, the spatiotemporal records of a user in datasets $\mathcal{X}$ and $\mathcal{Y}$ are assumed to be generated by an unknown but consistent probabilistic distribution.
- **Exclusiveness.** In each dataset, any two records with distinct anonymized identifiers and spatiotemporal attributes is exclusively generated by two different users. That is, no two records are produced by one user.

On the one hand, the *consistency* assumption ensures that the users across two datasets are identifiable based on their spatiotemporal behaviors. On the other hand, the *exclusiveness* assumption provides critical information on user matching or user identification such that user matching accuracy could be largely improved via a minimum-cost bipartite matching problem formulation discussed as follows [1].

7.2.2   Formulation: k-Cardinality Bipartite Matching

In datasets $\mathcal{X}$ and $\mathcal{Y}$, we assume that k users coexist with different anonymized identifiers in these two datasets, where $k = |\mathcal{X} \cap \mathcal{Y}|$, $|\mathcal{X}| = N$, and $\mathcal{Y} = M$. Hence, the objective of user identification in this work is to find a matching set with cardinality k,

$$\Phi^k = \{(i, j)|\sigma(i) = j, (i, X_i) \in \mathcal{X} \text{ and } (j, Y_j) \in \mathcal{Y}\}, \ |\Phi^k| = k \qquad (7.2)$$

such that X_i and Y_j are generated by the same user in these two datasets, where $\sigma(\cdot)$ denotes the permutation operator. With a distance measure on these two spatiotemporal attributes, the user identification problem could be formulated into *k-cardinality minimum-cost bipartite matching* (k-MinBM) based on the *exclusiveness* assumption discussed previously. The bipartite graph $\mathbb{G}(\mathcal{X}, \mathcal{Y}, \mathcal{E})$ is fully connected with vertices representing users in these two datasets as shown in Fig. 7.1. Edges are defined on any pair of two vertices respectively from two vertex sets, $\mathcal{E} = \{(i, j)|\forall(i, X_i) \in \mathcal{X} \text{ and } \forall(j, Y_j) \in \mathcal{Y}\}$, and their corresponding weights are defined as

$$w_{ij} = \Delta(X_i, Y_j), \qquad (7.3)$$

where $\Delta(X_i, Y_j)$ denotes a distance measure based on a specific modeling and feature extraction on spatiotemporal attributes, X_i and Y_j, which will be discussed thoroughly in Sect. 7.3. With definition of the bipartite graph $\mathbb{G}$, the k-MinBM problem takes the following form,

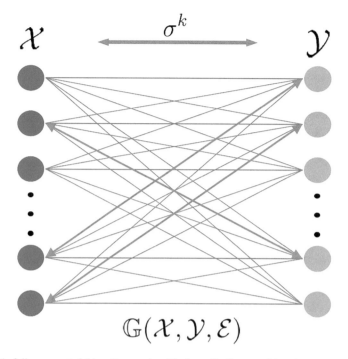

Fig. 7.1 A fully-connected bipartite graph with k-cardinality matching (orange arrow lines), $\mathbb{G}(\mathcal{X}, \mathcal{Y}, \mathcal{E})$, where vertex set denote the user set from dataset $\mathcal{X}$ and $\mathcal{Y}$, respectively. Each edge weight between two user i and j is obtained by a careful designed distance measure on their associated spatiotemporal attributes, $w_{ij} = \Delta(X_i, Y_j)$

$$
\begin{aligned}
\underset{c_{ij}}{\text{minimize}} \quad & \sum_{i}^{N}\sum_{j}^{M} c_{ij} w_{ij} \\
\text{subject to} \quad & \sum_{j}^{M} c_{ij} \leq 1, \forall i \in [N] \\
& \sum_{i}^{N} c_{ij} \leq 1, \forall j \in [M] \\
& \sum_{i}^{N}\sum_{j}^{M} c_{ij} = k, \ c_{ij} \in \{0, 1\}
\end{aligned}
\tag{7.4}
$$

In the literature, Hungarian algorithm is the classic solution to the MinBM problem [12], and its extension of the k-cardinality MinBM problem has been discussed in [13, 14].

7.2.3 Key Challenges

With the k-MinBM formulated, two critical problems of the studied user identification problem are mainly discussed as follows.

- **Representation and Distance Measures.** k-MinBM could be regarded as a deterministic procedure to obtain a matching after weights are generated according to a distance measure. Hence, the distance measure (7.3) is the key towards capturing the distance between two spatiotemporal attributes in datasets $\mathcal{X}$ and $\mathcal{Y}$. In fact, a distance measure is defined on one feature extracted from the spatiotemporal attribute, which may provide a narrow description of users' spatiotemporal behaviors. Good distance measures should be able to well distinguish the pair of spatiotemporal attributes generated by the same user from that by different users in the context of complex spatiotemporal behaviors of human being. In this case study, we first explore several semantic spatiotemporal features, based on which proposed distance measures could provide the quantity to assess the distance between two spatiotemporal attributes.
- **Ensemble Matching.** Based on each semantic spatiotemporal feature and its corresponding distance measure, the k-MinBM could produce a matching result based on the minimum sum distance criterion. However, different distance measures will give rise to different matching results, each of which may contain many false matched pairs. Here, we study an ensemble approach to effectively integrate multiple matching results based on diverse semantic spatiotemporal features so that the overall performance could be significantly improved.

As a result, we propose a *multi-feature ensemble matching framework* for the studied user identification problem (as shown in Fig. 7.2), in which the two main

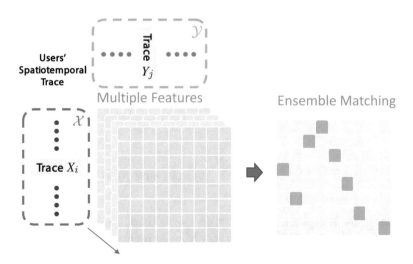

Fig. 7.2 The proposed multi-feature ensemble matching user identification framework

components will be thoroughly discussed in Sects. 7.3 and 7.4, respectively. In addition, the number of coexisting users across two datasets, k, is generally unknown. Hence, the assumed foreknowledge of k in [1, 2] is not practical in many scenarios, and the most intuitive approach is to set k to be as large as possible, i.e., $k = \min(N, M)$. However, such an approach will introduce a great number of false matches, especially when the coexisting number is significantly smaller than $\min(N, M)$. In this case study, we will show that the studied multi-feature ensemble matching framework could markedly reduce false matches *without* an explicit k estimation.

7.3   Representations and Distance Measures

As stated previously, distance measures play a critical role in user identification, which could largely determine the performance of a user identification algorithm based on the k-MinBM problem formulation. However, the raw spatiotemporal attributes could not be directly employed to assess the distance between two spatiotemporal attributes without a proper data modeling and feature extraction. In fact, each feature extracted from the raw spatiotemporal attributes could be regarded as a fingerprint of the user. In this section, we will discuss three semantic spatiotemporal features and their corresponding distance measures, namely location visiting frequency [1, 2], location-dependent dwelling time, and daily habitat regions, as shown in Fig. 7.3. The latter two spatiotemporal features are to profile users' spatiotemporal behaviors in different aspects, which could contribute significantly to the user identification task. Each feature and its corresponding distance measures will be discussed in terms of *data modeling*, *representing feature*, and the corresponding *distance measures*. It is also worth noting that each feature may have more than one distance measures.

7.3.1   Location Visiting Frequency Modeling [1, 2]

Modeling, Representation and Distance Measures

The location visiting frequency is utilized as a representing feature [1, 2] to distinctly characterize a user.

Data Modeling
    With the location point set (base station set) being abstracted as an alphabet set $\mathcal{A} = \{a_1, \cdots . a_l\}$, the raw spatiotemporal attribute (7.1) could be first modeled as a string with length T by discarding time information,

$$X_i = x_{i1}, x_{i2}, \cdots , x_{iT}. \tag{7.5}$$

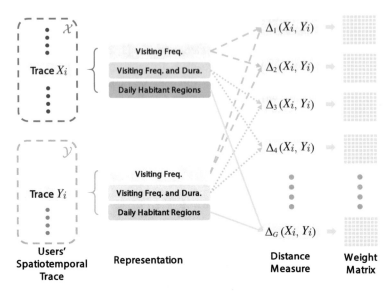

Fig. 7.3 Representation and distance measures

Every element $x_{it} \in \mathcal{A}$ in the string is assumed to be i.i.d. from the alphabet set $\mathcal{A}$ based on an unknown location visiting probability mass function Π_i.

Representing Feature

Based on the i.i.d assumption of the string generation, the location visiting probability Π_i could be estimated by the empirical probability distribution or histogram Γ_{X_i}, i.e.,

$$\Gamma_{X_i}(a_l) = \frac{N_i(a_l)}{T}, a_l \in \mathcal{A}, \tag{7.6}$$

where $N_i(a_l) = \sum_{x_{it}=a_l} 1$ denotes the number of the letter a_l appearing in the string X_i, counting the number of visits of user U_i at location point a_l. Thus, the spatiotemporal behaviors of a user could be represented by the *histogram*, characterizing his/her *visiting frequency* over location point set $\mathcal{A}$. However, such feature extraction further discards the temporal information for model simplicity. An example of visiting frequency of two users in two datasets is shown in Fig. 7.4.

Distance Measures

Hence, to evaluate whether two spatiotemporal attributes are associated with the same user in terms of visiting frequency is to find a good distance measure to assess the distance between two histograms. Here, the intuitive yet heuristic L_1 distance function could be employed to assess the distance between two histograms as follows,

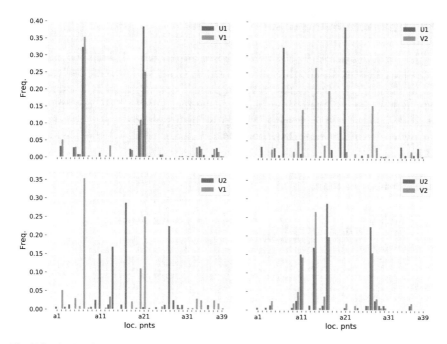

Fig. 7.4 Visiting frequency (empirical probability distribution) comparison between User 1 and User 2 in two datasets U and V, respectively

$$\Delta_{L_1\_f}(X_i, Y_j) = \sum_{a_l \in \mathcal{A}} \left| \Gamma_{X_i}(a_l) - \Gamma_{Y_j}(a_l) \right| . \tag{7.7}$$

Based on the multi-hypothesis test framework discussed in [1], to determine optimum hypothesis using the log likelihood test (7.11) is equivalent to solving k-MinBM with weights generated by the Jensen-Shannon divergence (JSdiv) on any pair of histograms. Thus, the JSdiv could be served as a distance measure on histograms defined as follows,

$$\Delta_{JS\_f}(X_i, Y_j) = \text{JSdiv}(\Gamma_{X_i}, \Gamma_{Y_j}) . \tag{7.8}$$

where

$$\text{JSdiv}(p, q) = \text{KL}\left(p \, \middle\| \, \frac{p+q}{2} \right) + \text{KL}\left(q \, \middle\| \, \frac{p+q}{2} \right) . \tag{7.9}$$

Details of (7.8) and (7.9) are discussed as follows.

Multi-Hypothesis Testing Framework [1]

In [1], a multi-hypothesis test framework based on generalized likelihood ratio test is utilized to guide the distance measure derivation. Given the knowledge of coexisting user number k, $\binom{n}{k}\binom{m}{k}k!$ possible matching candidates may need to be examined so that the one with optimum scores will be determined as the final matching. Under a multi-hypothesis test framework, each k-cardinality matching candidate is viewed as a hypothesis. A toy example of multiple hypothesis for 3-by-3 matching is shown in Fig. 7.5.

Thus, the generalized log likelihood of hypothesis $\mathcal{H}_r$ for the r-th matching candidate takes the form,

$$\mathcal{L}(\mathcal{H}_r) = \sup_{\Pi_{ij}, \Pi_{i'}, \Pi_{j'}} \sum_{(i,j) \in \Phi_r^k} \log\left(\Pi_{ij}(X_i)\right) + \log\left(\Pi_{ij}(Y_j)\right)$$

$$+ \sum_{(i',j') \notin \Phi_r^k} \log(\Pi_{i'}(X_{i'})) + \sum_{t=1}^{T} \log(\Pi_{j'}(Y_{j'})) \qquad (7.10)$$

where Π_{ij} denotes the unknown probability distribution based on the assumption that X_i and Y_j are generated by the same user in a matching candidate. And $\Pi_{i'}$ and $\Pi_{j'}$ are the unknown probability distribution of $X_{i'}$ and $Y_{j'}$, which are respectively assumed to be exclusive in their own datasets. Actually, Π_{ij}, $\Pi_{i'}$, and $\Pi_{j'}$ rely on the modeling of spatiotemporal attributes, and will be first estimated by maximum likelihood in the generalized log likelihood ratio test. The matching is finally determined by the one with maximum likelihood (scores), i.e.,

$$\max_r \mathcal{L}(\mathcal{H}_r) . \qquad (7.11)$$

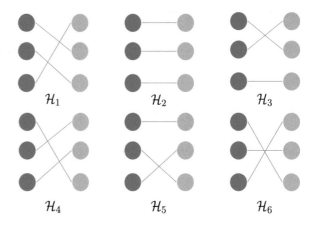

Fig. 7.5 A toy example of multiple-hypothesis test framework. As for $m = n = k = 3$, $\binom{3}{3}\binom{3}{3}3! = 6$ hypotheses could be formed, each of which represents a candidate 3-cardinality matching

Distance Measure by JS Divergence on Visiting Frequency

Thus, the log probability of string X_i under the probability distribution Π_i is

$$
\mathcal{L}(X_i) = \log\left(\prod_{t=1}^{T} \Pi_i(x_{it})\right) = \sum_{a \in \mathcal{A}} \log\left(\Pi_i(a)^{N_i(a)}\right)
$$

$$
= T \sum_{\underbrace{a \in \mathcal{A}}_{\text{negative cross entropy}}} \Gamma_{X_i}(a) \log\left(\Pi_i(a)\right) \; . \tag{7.12}
$$

The log likelihood of string X_i is essentially product of string length T and negative cross entropy between the empirical distribution Γ_{X_i} and the assumed background distribution Π_i, where cross entropy is defined as $H(p, q) = H(p) + KL(p\|q)$. with $H(\cdot)$ and $KL(\cdot)$ denoting the entropy and Kullback-Leibler (KL) divergence, respectively. Therefore, the log likelihood of hypothesis $\mathcal{H}_r$ (7.10) could be rewritten based on (7.12) as follows,

$$
\mathcal{L}(\mathcal{H}_r) = - T \sup_{\substack{\Pi_{ij} \\ (i,j) \in \Phi_r}} \underbrace{\sum_{X_i \in \mathcal{X}} H(\Gamma_{X_i}) + \sum_{Y_j \in \mathcal{Y}} H(\Gamma_{Y_j})}_{\text{empirical entropy}}
$$

$$
+ \sum_{(i,j) \in \Phi_r} \underbrace{KL(\Gamma_{X_i}\|\Pi_{ij}) + KL(\Gamma_{Y_j}\|\Pi_{ij})}_{\text{KL divergence}} \tag{7.13}
$$

Based on the strategy of the generalized log likelihood ratio test, the final matching result could be obtained by (7.11). For each hypothesis, the empirical entropy part in (7.13) could be omitted since its value is a constant for all hypotheses. In fact, to maximize $\mathcal{L}(\mathcal{H}_r)$ in (7.13) is essentially to minimize its corresponding KL divergence parts. Assumed that users are uncorrelated in terms of generating strings, the probability mass function generating both X_i and Y_j is estimated based on maximum likelihood estimation [1] as $\widehat{\Pi}_{ij} = \frac{1}{2}\left[\Gamma_{X_i} + \Gamma_{Y_j}\right]$. Under such generalized log-likelihood multi-hypothesis test formulation, it could be easily observed that the final matching could be obtained by minimizing the KL parts of (7.13), which is equivalent to perform the (7.4) with pair-wise weights calculated by (7.9). Thus, Jensen-Shannon divergence based distance measure on visiting frequency can be obtained (7.8).

7.3.2 *Location Visiting Frequency and Dwelling Time Modeling*

Modeling, Representation and Distance Measures

The previously discussed visiting frequency feature only captures one spatial aspect of the available spatiotemporal attributes, while neglecting the potential temporal information valuable for user identification. Though the collected dataset may be an event log with users' spatiotemporal trajectory sporadically sampled, the temporal information could still be utilized to feature the distinctness of users. In this subsection, we study a visiting frequency and duration feature to jointly evaluate the distance in terms of both the spatial and temporal aspects.

Data Modeling

Atop the previous string model (7.5), the raw spatiotemporal attribute (7.1) could be modeled as a tuple string with size P_i as follows,

$$X_i = (x_{i1}, t_{i1}), (x_{i2}, t_{i2}), \cdots , (x_{iP_i}, t_{iP_i}) , \tag{7.14}$$

where $x_{ip} \in \mathcal{A}$ denotes the p-th recorded location point of user i, and t_{ip} denotes its time length staring from the current event to next event. Here, each user may have unique string length.

Based on the spatiotemporal tuple string modeling, we also assume that each tuple is i.i.d. generated by an unknown probability distribution, where the duration of a user at a given location point $a_l \in \mathcal{A}$ is modeled as an exponential distribution condition on location a_l,

$$f(t|a_l; \lambda_{i,l}) = \lambda_{i,l} \exp(-\lambda_{i,l} t), \ t > 0, \tag{7.15}$$

where $\lambda_{i,l}$ denotes the reciprocal of average duration of user i at location point a_l.

Representing Features

Assumed that duration generated at locations are uncorrelated, the likelihood of X_i takes the form

$$\mathcal{L}(X_i) = \prod_l \mathcal{L}(X_i; a_l) \Pi_i(a_l) \tag{7.16}$$

where Π_i denotes the location frequency and $\mathcal{L}(X_i; a_l)$ denotes the likelihood of X_i observed at location a_l as follows,

$$\mathcal{L}(X_i; a_l) = \log \left[\prod_{z_{ip}=a_l} \lambda_{i,l} \exp\left(-\lambda_{i,l} t_{ip}\right) \right]$$

$$= N_i(a_l) \log(\lambda_{i,l}) - \lambda_{i,l} \sum_{z_{ip}=a_l} t_{ip}. \tag{7.17}$$

By taking the derivative of $\mathcal{L}(X_i; a_l)$ with respect to $\lambda_{i,l}$, the maximum likelihood estimate of $\lambda_{i,l}$ could be obtained by solving the equation that the derivative is equal to zero, which is essentially the reciprocal of the average of observation appearing in X_i at each a_l,

$$\hat{\lambda}_{i,l} = N_l(a_l) / \sum_{x_{ip}=a_l} t_{ip}. \tag{7.18}$$

As a result, a user could be represented as both the estimated visiting frequency histogram Γ_{X_i} by (7.6) and the duration model parameter set $\widehat{\Lambda}_i = \{\widehat{\lambda}_{i,l}\}$ from the spatial and temporal aspects, respectively (Fig. 7.6).

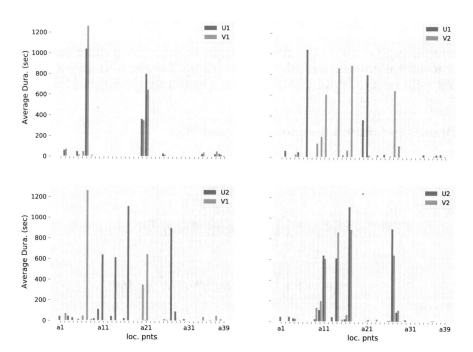

Fig. 7.6 Per-location average duration comparisons ($\Gamma_{X_i}(a_l)/\hat{\lambda}_{i,l}$ and $\Gamma_{Y_j}(a_l)/\hat{\lambda}_{j,l}$) between User 1 and User 2 in two datasets U and V, respectively

Distance Measures

With the multi-hypothesis test framework (7.10) and (7.11), the distance measure between two users in terms of both visiting frequency and location-dependent duration is obtained as follows,

$$\Delta_{\text{wdiv\_fd}}(X_i, Y_j) = \Delta_{\text{wdiv\_f}}(X_i, Y_j) + \Delta_{\text{wdiv\_d}}(X_i, Y_j) \tag{7.19}$$

where $\Delta_{\text{wdiv\_f}}(X_i, Y_j) = \text{wdiv}_{q_i}(\Gamma_{X_i}, \Gamma_{Y_j})$ is the distance measure originated from the visiting frequency modeling and the distance measure on the location-dependent duration modeling is

$$\Delta_{\text{wdiv\_d}}(X_i, Y_j) = \sum_{a_l \in \mathcal{A}} \left[q_i \Gamma_{X_i}(a_l) \text{KL}\left(\hat{\lambda}_{i,l} \| \hat{\lambda}_{ij,l} \right) \right. \\ \left. + q_j \Gamma_{Y_j}(a_l) \text{KL}\left(\hat{\lambda}_{j,l} \| \hat{\lambda}_{ij,l} \right) \right] \tag{7.20}$$

where $\text{KL}(\lambda_1 \| \lambda_2)$ denotes the KL divergence on two exponential distributions, i.e., $\text{KL}(\lambda_1 \| \lambda_2) = \log(\lambda_1/\lambda_2) + (\lambda_2/\lambda_1) - 1$. $\text{wdiv}(\cdot)$ is the weighted divergence,

$$\text{wdiv}_\omega(p, q) = \omega \text{KL}(p \| \omega(p, q)) + (1 - \omega)\text{KL}(q \| \omega(p, q)),$$

where $\omega(p, q) = \omega p + (1 - \omega)q$. The weighted divergence is a generalization of Jensen-Shannon divergence, which replaces $1/2$ in JSDiv with ω, where $\omega \in [0, 1]$. The weight ω in (7.19) and (7.20) is obtained based on asymmetric string lengths.

Derivation of Distance Measures (7.19) and (7.20)

The likelihood that two tuple strings X_i and Y_j by (7.14) takes the form based on the assumption that both are generated by the same user as follows,

$$\mathcal{L}(X_i, Y_j | \Pi_{ij}, \Lambda_{ij}) = \prod_{p=1}^{P_i} f(t_{ip} | x_{ip} = a_l; \lambda_{ij,l}) \Pi_{ij}(x_{ip}) \\ \times \prod_{p=1}^{P_j} f(t_{jp} | y_{jp} = a_l; \lambda_{ij,l}) \Pi_{ij}(y_{jp}) \tag{7.21}$$

where Λ_{ij} denotes the collection of $\lambda_{ij,l}$. Based on the exponential distribution assumption of duration observation conditioned on each location point (7.17), the logarithm version of (7.21) could be further studied as follows,

$$\mathcal{L}(X_i, Y_j | \Pi_{ij}, \Lambda_{ij})$$

$$= \sum_{a_l \in \mathcal{A}} N_i(a_l) \left[\log(\Pi_{ij}(a_l)) + \log \lambda_{ij,l} - \lambda_{ij,l} \bar{t}_{i,l} \right]$$

$$+ N_j(a_l) \left[\log(\Pi_{ij}(a_l)) + \log \lambda_{ij,l} - \lambda_{ij,l} \bar{t}_{j,l} \right] ,$$

where $\bar{t}_{i,l} = \frac{\sum_{p, z_{ip} = a_l} t_{ip}}{N_i(a_l)}$ and $\bar{t}_{j,l} = \frac{\sum_{p, z_{jp} = a_l} t_{jp}}{N_j(a_l)}$ denote the average time length of users i and j at location point a_l, respectively. The problem of maximum log likelihood (7.3.2) takes the form,

$$\underset{\Pi_{ij}, \Lambda_{ij}}{\text{maximize}} \quad \sum_{a_l \in \mathcal{A}} [N_i(a_l) + N_j(a_l)][\log \Pi_{ij}(a_l) + \log \lambda_{ij,l}]$$

$$- \lambda_{ij,l}[N_i(a_l)\bar{t}_{i,l} + N_j(a_l)\bar{t}_{j,l}] \tag{7.22}$$

$$\text{subject to} \quad \sum_{a_l \in \mathcal{A}} \Pi_{ij}(a_l) = 1$$

It could be observed that the empirical probability distribution Π_{ij} is independent from the exponential distribution of duration for any given location points. As a result, the estimate of Π_{ij} at each location point takes the form,

$$\widehat{\Pi}_{ij}(a_l) = q_i \Gamma_{X_i} + q_j \Gamma_{Y_j} = \frac{N_{ij}(a_l)}{P_{ij}}, \quad \forall a_l \in \mathcal{A}, \tag{7.23}$$

where $q_i = P_i / P_{ij}$, $q_j = P_j / P_{ij}$, $P_{ij} = P_i + P_j$, and $N_{ij}(a_l) = N_i(a_l) + N_j(a_l)$. Furthermore, the corresponding estimate of $\lambda_{ij,l}$ at each location point could be obtained as follows,

$$\widehat{\lambda}_{ij,l} = \frac{1}{k_{i,l}/\widehat{\lambda}_{i,l} + k_{j,l}/\widehat{\lambda}_{j,l}} = \frac{N_{ij}(a_l)}{\tilde{t}_{i,l} + \tilde{t}_{j,l}}, \quad \forall a_l \in \mathcal{A}, \tag{7.24}$$

where $\tilde{t}_{i,l} = \sum_{p, z_{ip} = a_l} t_{ip}$ and $\tilde{t}_{j,l} = \sum_{p, z_{jp} = a_l} t_{jp}$ denote the sum of durations at a_l of users i and j, respectively. $k_{i,l} = N_i(a_l)/N_{ij}(a_l)$ and $k_{j,l} = N_j(a_l)/N_{ij}(a_l)$ denote the weights.

Here, $\hat{\lambda}_{i,l}$ and $\hat{\lambda}_{j,l}$ are maximum likelihood estimates of X_i and Y_j, respectively. Based on the multi-hypothesis test framework (7.10), the log likelihood of hypothesis $\mathcal{H}_r$ could be expressed as follows,

$$\mathcal{L}(\mathcal{H}_r) = \sup_{\substack{\Pi_{ij}, \Pi_i, \Pi_j, \\ \Lambda_{ij}, \Lambda_i, \Lambda_j}} \sum_{(i,j) \in \Phi_r} \log \left[\mathcal{L}(X_i, Y_j | \Pi_{ij}, \Lambda_{ij}) \right]$$

$$+ \sum_{(i,j) \notin \Phi_r} \log \left[\mathcal{L}(X_i | \Pi_i, \Lambda_i) \right] + \log \left[\mathcal{L}(Y_j | \Pi_j, \Lambda_j) \right] \tag{7.25}$$

where $\mathcal{L}(X_i, Y_j | \Pi_{ij}, \Lambda_{ij}) = \mathcal{L}(X_i | \Pi_{ij}, \Lambda_{ij}) \mathcal{L}(Y_j | \Pi_{ij}, \Lambda_{ij})$ based on the i.i.d. assumption.

The likelihood function (7.21) could be rewritten in terms of two components, namely visiting frequency and location-dependent duration as follows,

$$-\frac{\mathcal{L}(X_i, Y_j | \Pi_{ij}, \Lambda_{ij})}{P_{ij}} = \mathcal{L}_{\text{freq}}(\Gamma_{X_i}, \Gamma_{Y_j}) + \mathcal{L}_{\text{dura}}(\Gamma_{X_i}, \Gamma_{Y_j}, \hat{\Lambda}_{ij}). \tag{7.26}$$

The first part, $\mathcal{L}_{\text{freq}}$, can be obtained on the frequency features by generalizing (7.13) in terms of unequal string lengths,

$$\mathcal{L}_{\text{freq}}(\Gamma_{X_i}, \Gamma_{Y_j}) = q_i H(\Gamma_{X_i}) + q_j H(\Gamma_{Y_j}) + \text{wdiv}_{q_i}(\Gamma_{X_i} \| \Gamma_{Y_j}). \tag{7.27}$$

The second component $\mathcal{L}_{\text{dura}}$ is related to duration modeling, representing the weighted sum of the cross entropy between the exponential distribution at each location point a_l. With parameter estimates $\widehat{\Pi}_{ij}$ and $\widehat{\Lambda}_{ij}$, $\mathcal{L}_{\text{dura}}(\widehat{\Pi}_{ij}, \widehat{\Lambda}_{ij})$ could be easily obtained based on (7.27) and (7.3.2) as follows,

$$\mathcal{L}_{\text{dura}}(X_i, Y_j | \widehat{\Pi}_{ij}, \widehat{\Lambda}_{ij}) = \sum_{a_l \in \mathcal{A}} \widehat{\Pi}_{ij}(a_l) \left(1 - \log \hat{\lambda}_{ij,l}\right). \tag{7.28}$$

With $\frac{1}{\hat{\lambda}_{ij,l}} = \frac{k_{i,l}}{\hat{\lambda}_{i,l}} + \frac{k_{j,l}}{\hat{\lambda}_{j,l}}$, $\mathcal{L}_{\text{dura}}$ could be further expressed at each a_l as follows,

$$\frac{\mathcal{L}_{\text{dura}}(a_l)}{\widehat{\Pi}_{ij}(a_l)} = k_{i,l} \left[\frac{\hat{\lambda}_{ij,l}}{\hat{\lambda}_{i,l}} - \log \hat{\lambda}_{ij,l} \right]$$
$$+ k_{j,l} \left[\frac{\hat{\lambda}_{ij,l}}{\hat{\lambda}_{j,l}} - \log \hat{\lambda}_{ij,l} \right]$$

The differential entropy of exponential distributions is $H(\lambda) = 1 - \log \lambda$. The KL divergence between two exponential distributions, λ_1 and λ_2, is $\text{KL}(\lambda_1 \| \lambda_2) = \log(\lambda_1/\lambda_2) + \lambda_2/\lambda_1 - 1$. Therefore, $\mathcal{L}_{\text{dura}}$ could be further rewritten in terms of entropies and KL divergences as follows,

$$\mathcal{L}_{\text{dura}} = \sum_{a_l \in \mathcal{A}} \widehat{\Pi}_{ij}(a_l) \left\{ k_{i,l} \left[H(\hat{\lambda}_{i,l}) + \text{KL}(\hat{\lambda}_{i,l} \| \hat{\lambda}_{ij,l}) \right] \right.$$
$$\left. + k_{j,l} \left[H(\hat{\lambda}_{j,l}) + \text{KL}(\hat{\lambda}_{j,l} \| \hat{\lambda}_{ij,l}) \right] \right\}$$

With $\widehat{\Pi}_{ij}(a_l) k_{i,l} = q_i \Gamma_{X_i}(a_l)$ and $\widehat{\Pi}_{ij}(a_l) k_{j,l} = q_j \Gamma_{Y_j}(a_l)$, $\mathcal{L}_{\text{dura}}$ could be further expressed as follows,

$$\mathcal{L}_{\text{dura}}(\Gamma_{X_i}, \Gamma_{Y_j}, \widehat{\Lambda}_i, \widehat{\Lambda}_j)$$

$$= \sum_{a_l \in \mathcal{A}} q_i \Gamma_{X_i}(a_l) \left[H(\hat{\lambda}_{i,l}) + \text{KL}(\hat{\lambda}_{i,l} \| \hat{\lambda}_{ij,l}) \right] \tag{7.29}$$

$$+ q_j \Gamma_{Y_j}(a_l) \left[H(\hat{\lambda}_{j,l}) + \text{KL}(\hat{\lambda}_{j,l} \| \hat{\lambda}_{ij,l}) \right]$$

Based on the similar reasoning as in (7.13), the entropy part of (7.27) and (7.29) could be eliminated for it is a constant for all the hypotheses. To determine the most likely hypothesis is to perform a k-cardinality minimum cost bipartite matching, where the edge weights are the pair-wise distance measure via both frequency and duration modeling. Thus, (7.19) and (7.20) could be easily obtained by keeping the divergence parts in (7.27) and (7.29).

As the large area is usually covered by the mobile network, almost no one could visit every base stations of the mobile network. As a result, two empirical distributions would have asymmetric probability supports, e.g., $\Gamma_{X_i}(a_l) = 0$, $\Gamma_{Y_j}(a_l) > 0$, $\exists a_l \in \mathcal{A}$, which may produce infinity value by the distance measure due to the asymmetric probability supports. Such property is not obvious in the distance measure on the location-dependent duration modeling as shown in (7.19) and (7.20). Here, we first rewrite (7.19) at location point a_l as follows,[1]

$$\Delta_{\text{wdiv\_fd}}(X_i, Y_j; a_l)$$

$$= \frac{N_i}{P_{ij}} \log\left(\frac{P_{ij}}{P_i}\right) + \frac{2N_i}{P_{ij}} \log \frac{N_i}{N_{ij}} + \frac{N_i}{P_{ij}} \log\left(\frac{\tilde{t}_{ij}}{\tilde{t}_i}\right) \tag{7.30}$$

$$+ \frac{N_j}{P_{ij}} \log\left(\frac{P_{ij}}{P_j}\right) + \frac{2N_j}{P_{ij}} \log \frac{N_j}{N_{ij}} + \frac{N_j}{P_{ij}} \log\left(\frac{\tilde{t}_{ij}}{\tilde{t}_j}\right)$$

Intuitively, the value of a distance measure on two tuple strings should be as small as possible when these two are generated by the same user; otherwise, it should be as large as possible so that the two users could be distinguished when their supports are asymmetric. Thus, we examine some boundary cases below:

1. When a_l is not observed in X_i but observed in Y_j, i.e., $N_i = 0$ and $\tilde{t}_i = 0$, the distance measure at a_l is

$$\Delta_{\text{wdiv\_fd}}(X_i, Y_j; a_l) = \frac{N_j}{P_{ij}} \log\left(\frac{P_{ij}}{P_j}\right) .$$

2. When a_l is not observed in X_i yet observed in Y_j, $\tilde{t}_i$ is assumed to exactly the same as $\tilde{t}_j$, i.e., $\tilde{t}_i = \tilde{t}_j$ and $N_i = 0$, then the distance measure at a_l is

---

[1]For notation simplicity, we get rid of notation a_l and subscript l in (7.30).

$$\Delta_{\text{wdiv\_fd}}(X_i, Y_j; a_l) = \frac{N_j}{P_{ij}} \log\left(\frac{P_{ij}}{P_j}\right) + \frac{N_j}{P_{ij}} \log 2 \ .$$

In this work, we choose (2) for asymmetric support distance measure calculation, since its value is larger than that of (1).

7.3.3   Geospatial Habitat Region Modeling

The previously discussed spatiotemporal features abstract discrete location points as independent and unrelated letters in an alphabet set $\mathcal{A}$. Such modeling discards the critical geo-spatial information, which generally describes the relationship between location points by the raw latitude and longitude coordinates. The geospatial information may help resist the information loss due to the sporadic sampling of users' spatiotemporal trajectories. Thus, we study a heuristic spatiotemporal feature for user identification, daily habitat regions, as well as its corresponding distance measure, based on the geospatial information in this subsection. The daily habitat regions capture the daily spatial coverage of a subscriber, which are expected to be consistent to some degree and may serve as subscriber's mobility fingerprints.

Data Modeling
    The spatiotemporal attribute (7.1) is first formulated into sets of location points, i.e.,

$$X_i = \{\mathcal{X}_{i1}, \mathcal{X}_{i2}, \cdots, \mathcal{X}_{iQ_X}\}, \tag{7.31}$$

where each set $\mathcal{X}_{iq} \subseteq \mathcal{A}$ denotes a set of location points that the user visits during calendar date q. By the assumption that two datasets may have different data collection time period lengths, where Q_X and Q_Y denote the number of days collected in dataset $\mathcal{X}$ and $\mathcal{Y}$, respectively.

Representing Feature
    Here, we employ a classical computational geometry concept, convex hull, to approximate the spatial coverage that a user visits daily. By approximating a small region of geo-surface as a Euclidean space, the convex hull of a given point set $\mathcal{X}_{iq}$ in a 2-dimensional surface is defined as the set of the convex combination of the given finite point set as follows,

$$C_{iq} = \left\{ \sum_{l=1}^{|\mathcal{X}_{iq}|} \beta_l a_l \ \middle| \ \forall l, \beta_l > 0, a_l \in \mathcal{X}_{iq} \ \text{ and } \ \sum_{l=1}^{|\mathcal{X}_{iq}|} \beta_l = 1 \right\} \ .$$

Thus, the daily convex hull, C_{iq}, is employed to represent the spatiotemporal behaviors of a user for a given day. Hence, the spatiotemporal attributes of user i is represented as a set of daily convex hulls,

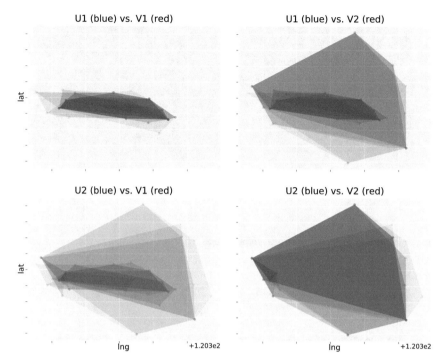

Fig. 7.7 Daily habitat regions (convex hulls) comparisons during data recorded time periods between User 1 and User 2 in two datasets U and V, respectively

$$[C_{i1}, C_{i2}, \cdots, C_{iQ_{X_i}}], \tag{7.32}$$

where each convex hull is again assumed to be i.i.d. generated by an unknown probability distribution (Fig. 7.7).

Distance Measure

With the convex hull set representing users' spatiotemporal behaviors, we first define a distance measure on two convex hulls based on the cosine distance between two polygons in terms of their overlapping area as follows,

$$\delta(C_p, C_q) = 1 - \frac{\text{area}(C_p \wedge C_q)}{\sqrt{\text{area}(C_p) \times \text{area}(C_q)}}, \tag{7.33}$$

where $C_p \wedge C_q$ denotes the overlapping region of the two convex hulls, and the operator area($\cdot$) is to calculate the area of a polygon. Therefore, a distance measure between two convex hull sets is studied based on (7.33) to evaluate the similarity of two subscribers as follows,

Features	Data Modeling	Representation	Distance Meas.
Visiting Frequency	Location String	Visiting Histogram	L_1 Distance JS Divergence
Visiting Freq. and Dura.	Location and Duration Tuple String	1. Visiting Histogram 2. Reciprocal of Average Duration Sets	Weighted Divergence
Daily Habitat Region	Daily Visiting Location Point Set	Convex Hull Set	Cosine Distance

Fig. 7.8 Summary of studied spatiotemporal features

$$\Delta_{\cos\_hr}(X_i, Y_j) = \frac{1}{Q_{X_i} \times Q_{Y_j}} \sum_{C_{ip} \in X_i} \sum_{C_{iq} \in Y_j} \delta(C_{ip}, C_{iq}) . \qquad (7.34)$$

Intuitively, the distance measure between two convex hull sets is to calculate the average distance between any two convex hulls in two respective sets. When the convex hull is not able to be obtained because the number of distinct visited location points within a day is less than 3, the daily habitat region would be omitted. If not a convex hull could be generated, the user will be labeled as non-identifiable. The studied spatiotemporal features are summarized in Fig. 7.8.

7.4 Ensemble Matching

As discussed previously, we have extracted three semantic spatiotemporal features, namely frequency, duration, and daily habitat regions. Each feature has at least one distance measure (summarized in Fig. 7.8), each of which could produce a matching result by solving k-MinBM (7.4) when k is assumed to be known, as shown in Fig. 7.2. In this section, we will discuss and explore an ensemble matching framework, which could effectively integrate results generated by multiple distance measures so that false matched pairs of the final matching could be largely eliminated without an explicit k estimation.

Ensemble learning is a category of algorithms to integrate multiple *weak learners* to obtain a much more powerful learner. Ensemble learning is originally designed for classification problem, where *weak learners* should satisfy following two criteria: (1) weak learners should be accurate to some degree (at least better than random guessing), which prevent the weak learner from contaminating final results;

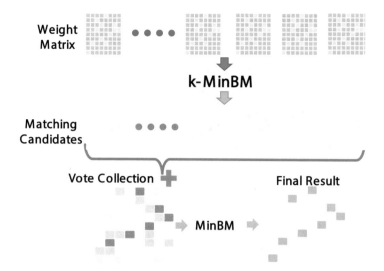

Fig. 7.9 Ensemble matching framework

(2) weak learners should be also diversified so that learners could capture different aspects rather than producing similar results that may make the ensemble fail.

In the literature, three types of ensemble learning are usually utilized, namely boosting, bagging, and stacking [11]. Here, the matching based on different distance measures can be regarded as a weak learner. Since the studied distance measures are originated from different semantic spatiotemporal features, the diversity requirement of ensemble learning is fulfilled. It could be also observed in experiments the matching result produced by each previously discussed distance measure along can capture the pairs generated by the same user albeit with many false matched pairs.

Hence, we investigate an ensemble matching framework to integrate/ensemble multiple results produced by diverse spatiotemporal feature based distance measures. None of existing frameworks could be directly applied, while the studied user identification problem is an unsupervised learning problem. However, the information fusion philosophy behind the "stacking" method inspires the studied ensemble matching. In addition, the exclusiveness property in our studied matching problem should be also enforced after the ensemble of multiple matching results. To integrate multiple distance measures we discussed previously, the naive approach is the weighted summation of distance measures before matching so that the exclusiveness could be guaranteed. However, without any proper training, directly distance measures and then applying the k-MinBM may lead to a even worse performance.

Instead of directly integrating the distance measures before solving k-MinBM, we ensemble the matching results produced by the k-MinBM based on diverse distance measures, as shown in Fig. 7.9. The matching based on a distance measure can be regarded as a filter to select k matched candidates from its own perspective out of massive $\binom{N}{k}\binom{M}{k}k!$ possibilities. With total G distance measures, let matrix $C^{(g,k)} \in \{0,1\}^{N \times M}$ denotes the matching result based on the g-th distance measure with the assumption of k coexisting user number, where its element takes the form as follows,

$$c_{ij}^{(g,k)} = \begin{cases} 1 & \sigma^{(g,k)}(i) = j \\ 0 & \text{otherwise} \end{cases}.$$

Let matrix $V^k \in \mathcal{Z}^{N \times M}$ collect the matching results by total G distance measures on each possible matching pair (i, j) with the assumption of k coexisting user number, i.e.,

$$V^{(k)} = \sum_{g=1}^{G} C^{(g,k)}. \tag{7.35}$$

Therefore, by the strategy of majority votes, the proposed ensemble matching is to solve following combinatoric optimization problem,

$$\begin{aligned} \underset{c_{ij}^{(F,k)}}{\text{maximize}} \quad & \sum_{i}^{N} \sum_{j}^{M} c_{ij}^{(F,k)} v_{ij}^{(k)} \\ \text{subject to} \quad & \sum_{j}^{M} c_{ij}^{(F,k)} \leq 1, \forall i \in [N] \\ & \sum_{i}^{N} c_{ij}^{(F,k)} \leq 1, \forall j \in [M], \ c_{ij}^{(F,k)} \in \{0,1\} \\ & c_{ij}^{(F,k)} (v_{ij}^{(k)} - \tau) \geq 0, \forall i \in [N], j \in [M] \end{aligned} \tag{7.36}$$

where $C^{(F,k)} \in \{0,1\}^{N \times M}$ denotes the final result generated by the studied ensemble matching framework with the assumption of k coexisting users. The first two conditions in (7.36) are exactly the same as the ones in (7.4), which guarantee the exclusiveness property. The τ denotes the threshold that ensures that the final result are produced based on majority votes, whose typical value is $\tau = \lceil G/2 \rceil$. As a result, the third condition, $c_{ij}^{(F,k)}(v_{ij}^{(F,k)} - \tau) \geq 0$, is the one that enforces the solution to (7.36) to be voted by majority. The objective function in (7.36) is aimed to maximize total votes generated by multiple distance measures without any explicit restriction on the cardinality of final results, as the cardinality restriction condition has already been enforced in (7.4) before ensemble matching.

In fact, the intuition behind vote maximization of (7.36) is to choose the one with more votes when the selection of both two candidate pairs violates the exclusiveness property, e.g., $\max(v_{ij}, v_{il})$, $v_{ij} \geq \tau$, , $v_{il} \geq \tau$. To solve the ensemble matching problem, we reformulate (7.36) into a classical minimum-cost bipartite matching problem as follows,

$$
\begin{aligned}
\underset{c_{ij}^{(F,k)}}{\text{minimize}} \quad & \sum_{i}^{n} \sum_{j}^{m} c_{ij}^{(F,k)} [G - \max(v_{ij}^{(k)} + 1 - \tau, 0)] \\
\text{subject to} \quad & \sum_{j}^{M} c_{ij}^{(F,k)} = 1, \forall i \in [N] \\
& \sum_{i}^{N} c_{ij}^{(F,k)} \leq 1, \forall j \in [M], \ c_{ij}^{(F,k)} \in \{0, 1\}
\end{aligned}
\tag{7.37}
$$

Without the loss of generality, we assume $N \leq M$ in (7.37). By the classical Hungarian algorithm, N pairs is generated, from which final results are determined by removing the matched pairs whose votes do not satisfy $v_{ij}^{(k)} \geq \tau$. Details of the studied ensemble matching framework are demonstrated in Fig. 7.9.

In summary, based on diverse distance measures from various aspects of users' spatiotemporal behaviors, the k-cardinality minimum cost bipartite matching could be regarded as a filter to largely reduce impossible matching pair candidates, and the studied ensemble matching framework as demonstrated in (7.36) is to integrate diverse matching results by ensuring both the exclusiveness property and the strategy of majority votes.

7.5 Experiments

In this section, we validate the studied feature extraction, distance measures and ensemble matching via experiments on a real-world signaling dataset collected in a mobile network, which is an extension of the commonly studied call detail records (CDR) dataset.

7.5.1 Signaling Dataset

The same signaling data as in the previous case study is utilized in this case study. Data fields of the signaling data include (1) *subscriber's anonymized identifier*, (2) *time stamp* (e.g., 20160101184312), (3) *location coordinates* (i.e., the longitude and latitude of the base station), (4) *event type*, and (5) *cell type* (i.e., small cell or macro cell). The longitude and latitude coordinates where the base station of each cell is located are accurate to six decimal places and time stamps are accurate to seconds. The signaling data logs event type as well as the direction of the event (e.g., initiating a call or being called).

Compared with the commonly used call detail records (CDR) datasets in the literature, the signaling data further logs two types of location update events besides the regular event types (calls or texts), namely regular location update and periodic location update. Location updating is an approach that the mobile network operator can learn the location of an inactive device, to which the call or text could be directed. The regular location update is triggered by a subscriber crossing a location area (in 3G) or a tracking area (in LTE), which cover much larger than a cell. The periodic location update is triggered by a timeout event that no event occurs for a subscriber within a predefined time interval, which is 1 h in the studied dataset. The periodic location updates in signaling data guarantee that any power-on subscriber of the mobile network has at least one observation within an hour in the dataset, compared with the commonly used CDR in the literature.

More than 6000 cells with millions of subscribers are recorded in the studied dataset. The number of daily recorded subscribers is about three million. The time period of the studied signaling data is 2 weeks, from January 1st, 2016 to January 14th, 2016. A small-size subscriber pool with total 6500 subscribers is created for experiments discussed later with three components: (1) 1000 subscribers are randomly selected; (2) Around 1000 subscribers are selected with conditions that they appear at least once in a region (residential area) from 12am to 5am; (3) 4500 subscribers are selected based on the condition that a subscriber appears in three non-overlapped regions in the daytime at least once.

7.5.2   Distance Measures

As discussed previously, the user identification performance is largely dependent on a good distance measure, as the dynamics and randomness root in users' spatiotemporal behaviors. Thus, the performance evaluation of distance measures before matching could be conducted based on the separation of a distance measure between two spatiotemporal attributes generated by the same user and the one generated by different users.

The normalized histograms of values generated by each distance measure summarized in Fig. 7.8 are demonstrated in Fig. 7.10. For each distance measure, two histograms on the value of the distance measure are computed, namely two spatiotemporal attributes generated by the same user (color red), and the ones generated by different users (color blue), where x axis records the value range of a distance measure, and y axis presents the normalized density.

In terms of the overlapped area of two histograms, all the distance measures generate a good separation. The distance measure $l1\_f$, i.e., applying L_1 distance function on visiting frequency, has the smallest overlapped area as shown in Fig. 7.10a, where the value of distance measure $l1\_f$ ranges from 0 to 2. The L_1 distance on visiting frequency also can achieve the best performance among

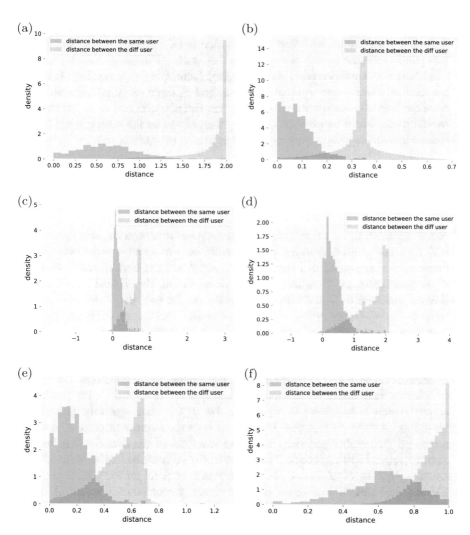

Fig. 7.10 Histogram of distance measures. (**a**) ll_f, (**b**) jsdiv_f, (**c**) jsdiv_d, (**d**) jsdiv_fd, (**e**) wdiv_fd, (**f**) cos_hr

matchers before ensemble matching, shown later in details. The distance measure JS divergence on visiting frequency jsdiv_f has a second smallest overlapped areas, where its value ranges from 0 to 0.69. It is worth noting we use the natural log here, so the maximum value generated by JSdiv_f is ln(2) ≈ 0.69.

It could be also observed that the overlapped area of two histograms in information theoretic distance measures (i.e., jsdiv and wdiv as shown in Fig. 7.10b–e) is located at the relative at left-hand side of x axis, which will make some pairs of spatiotemporal attributes respectively produced by two distinct users wrongly identified as the ones generated by the same user. On the other hand, the overlapped area of two histograms of distance measure cos_hr resides relatively at right-hand side of x axis, which will lead to the difficulty of the matcher to discover the true pairs generated by the same user. In fact, such phenomenon results from the curtailment of spatiotemporal details during habitat region modeling. However, the relatively worse performance of the studied distance measures does not make them useless. The relatively weaker cos_hr distance measure does improve the overall performance by providing an unique aspect under the studied multi-feature ensemble matching framework suggested as in next subsection.

7.5.3 User Identification Performance

A test scenario is set up to ensure $n = |\mathcal{X}| = 1000$ and $m = |\mathcal{Y}| = 1000$ with $k = 600$ coexisting users, so the performance of matchings for every experiment could be evaluated. Figure 7.11 demonstrates the performance evaluation of discussed distance measures for user identification. Experiment results are the average obtained by 200 randomly sampling on the user pool with the scenario setup enforced. Under such setting, two evaluation metrics are compared between matchings by toggling the declared coexisting user number k.

Figure 7.11a, b present the receiver operation characteristics (ROC) like curve of various distance measures without ensemble and with ensemble matching, respectively. The x axis in both figures is the number of false matches out of total matched pairs declared by distance measures, while the y axis is the number of correct matches. As shown in Fig. 7.11a, the ll_f is once again the best among all discussed distance measures in terms of both the number of correct identified users and the number of falsely identified users out of the declared matches. On average, around 550 users out of the ground truth 600 could be identified by distance measure ll_f. In other words, not all the spatiotemporal attribute pairs could be identifiable based on a distance measure. Such phenomenon is applicable to all distance measures, which might result from the inappropriate model assumption for some users or exactly the same spatiotemporal behaviors for different users (e.g., reside at one cell in the entire week).

It could be also observed that the studied joint visiting frequency and duration based distance measures can achieve slightly weaker performance, compared with the ll_f. The one based on heuristic daily habitat regions is the worst among all plotted ROC curves, for the distance measure based on daily habitat region is sensitive to user's spatiotemporal dynamics, e.g., if a specific user went to somewhere special out of the covered area that the user usually goes in 24 h, the habitat region in that day may be completely different. However, it still identifies almost half of true pairs (around 300 out of 600) as shown in Fig. 7.11a, regardless of false matches. The inferior performance of distance measure cos_hr independently does not make it useless. With the studied ensemble matching framework, it can contribute to user identification by providing a distinct characterization of user's spatiotemporal behaviors.

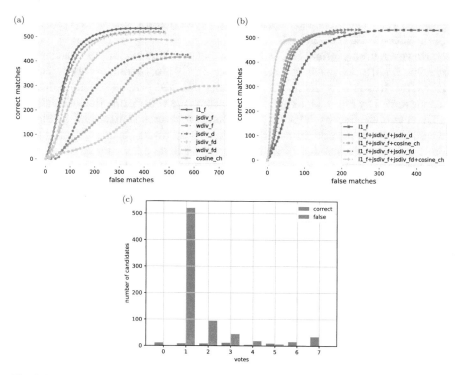

Fig. 7.11 Experiment results. The test dataset has about 6500 users with 2-week data including 1000 users randomly sampled from the entire dataset, about 1000 users residing in a region during the midnight, and around 4500 users visiting three predefined areas within 24 h. The scenario $n = m = 1000$ with $k = 600$ is tested, where n and m denote the user number during the first week and the second week, respectively. The tested users are randomly sampled from the test dataset with the fulfillment of scenario setup. The curves shown in the figures are the average of 200 random samplings. Legends denote the distance measure and features employed for matching, which are summarized in Fig. 7.8. (**a**) correct vs. false w/o ensemble, (**b**) correct vs. false w/ ensemble, (**c**) vote distribution

Figure 7.11b shows the performance of ensemble matching, compared with the superior one *l1_f* without ensemble matching. Curve legends in Fig. 7.11b suggest distance measures employed in each ensemble matching. It is obvious that the ensemble matching significantly improves the matching performance for user identification in terms of false match reduction. The results by integrating three distance measures as demonstrated in Fig. 7.11b indicate the contribution of distance measures, in which two distance measures, *l1_f* and *jsdiv_f*, are employed in all three ensembles. The performance of the three ensembles are similar to each other, but the one involved with daily-habitat-region-based distance measure is slightly the best among the three ensembles, as it provides a more diverse analysis of user's behavior than the other two, compared with the commonly used visiting-frequency-based distance measures. Although the total correctly matched pairs of the ensembles shrink compared with the best individual one *l1_f*, the false-to-declared ratio is significantly reduced (72.3% less) from 46.5% (l1_f) to 12.9% (4-distance-measure ensemble).

However, the performance gain by ensemble matching in terms of reducing false matches is not a free lunch, as it can achieve slightly less maximum correct matched pairs. The reason why the number of correct matched pairs is slightly reduced is demonstrated by Fig. 7.11c. Figure 7.11c records the vote distribution of candidates after vote collection by involving all distance measures summarized in Fig. 7.8, where the x axis records the number of votes by all diverse distance measures and the y axis logs the number of spatiotemporal attribute pairs corresponding to the number of votes. It can be observed that most of false matched candidates have votes less than majority, of which a large portion have only one vote. Hence, the majority-vote condition acts as a filter in (7.37), largely curtailing the false matches, but it may also trim a small part of correct matched as indicated by Fig. 7.11c. In addition, the tradeoff could be observed more obviously when more distance measures are integrated in ensemble matching.

7.6 Discussions and Summary

Overall, the subscriber privacy is vulnerable in terms of user identifiability across two datasets, if the dataset is released only with identifier anonymization. In the literature, to discover as many as possible correct pairs is the major objective without false matches considered. However, although correct matched pairs are included as many as possible in a declared matching, user's privacy could be still maintained to some extent if many false matched pairs also largely exist in the declared matching. In other words, correct matched pairs are hidden under false matched matches, especially when the number of coexisting user across two datasets are small. This is the reason why we intent to reduce false matches from the perspective of privacy attacker. As the studied ensemble matching framework relies on the diverse features extracted from data, detailed information reduction may help protecting user's privacy. For example, the daily habitat region based distance measure relies on the exact location coordinates of base stations, and the curtailment of location coordinate information would make such distance measure unavailable, which in turn reduces the performance of ensemble matching.

To sum up, we studied privacy attack in terms of user identifiability across two datasets based on spatiotemporal data collected from mobile networks. With k-cardinality minimum cost bipartite matching formulation, a multi-feature ensemble matching framework was studied. In this case study, we first studied to extract two new semantic spatiotemporal features as well as their associated distance measures. With multiple matching results via diverse features, an ensemble matching framework was studied to fuse matching results so that the final result is solid and robust. Experiments demonstrated the studied multi-feature ensemble matching achieved a superior performance (72.2% less false-to-declared ratio), which also suggested the vulnerability of mobile network subscriber's privacy.

References

1. J. Unnikrishnan, "Asymptotically optimal matching of multiple sequences to source distributions and training sequences," *IEEE Transactions on Information Theory*, vol. 61, no. 1, pp. 452–468, Jan. 2015.
2. F. M. Naini, J. Unnikrishnan, P. Thiran, and M. Vetterli, "Where you are is who you are: User identification by matching statistics," *IEEE Transactions on Information Forensics and Security*, vol. 11, no. 2, pp. 358–372, Feb. 2016.
3. X. Cheng, L. Fang, X. Hong, and L. Yang, "Exploiting mobile big data: Sources, features, and applications," *IEEE Network*, vol. 31, no. 1, pp. 72–79, January 2017.
4. Y. De Mulder, G. Danezis, L. Batina, and B. Preneel, "Identification via location-profiling in GSM networks," in *Proceedings of the 7th ACM Workshop on Privacy in the Electronic Society*, Alexandria, Virginia, USA, 2008, pp. 23–32.
5. Y.-A. de Montjoye, C. A. Hidalgo, M. Verleysen, and V. D. Blondel, "Unique in the crowd: The privacy bounds of human mobility," *Scientific Reports*, vol. 3, Mar. 2013.
6. A. Cecaj, M. Mamei, and N. Bicocchi, "Re-identification of anonymized CDR datasets using social network data," in *Proceedings of IEEE International Conference on Pervasive Computing and Communication Workshops (PERCOM WORKSHOPS)*, Budapest, Hungary, Mar. 24–28, 2014, pp. 237–242.
7. M. Gramaglia and M. Fiore, "Hiding mobile traffic fingerprints with GLOVE," in *Proceedings of the 11th ACM Conference on Emerging Networking Experiments and Technologies*, Heidelberg, Germany, Dec. 1–4, 2015, pp. 26:1–26:13.
8. M. Gramaglia, M. Fiore, A. Tarable, and A. Banchs, "Preserving mobile subscriber privacy in open datasets of spatiotemporal trajectories," in *Proceedings of IEEE International Conference on Computer Communications (INFOCOM)*, Atlanta, GA, USA, May 1–4, 2017, pp. 1–9.
9. B. C. M. Fung, K. Wang, R. Chen, and P. S. Yu, "Privacy-preserving data publishing: A survey of recent developments," *ACM Comput. Surv.*, vol. 42, no. 4, pp. 14:1–14:53, Jun. 2010.
10. C. Riederer, Y. Kim, A. Chaintreau, N. Korula, and S. Lattanzi, "Linking users across domains with location data: Theory and validation," in *Proceedings of the 25th International Conference on World Wide Web*, Montreal, Quebec, Canada, Apr. 11–15, 2016, pp. 707–719.
11. Z.-H. Zhou, *Ensemble methods: foundations and algorithms*. CRC press, 2012.
12. R. Jonker and T. Volgenant, "Improving the Hungarian assignment algorithm," *Operations Research Letters*, vol. 5, no. 4, pp. 171–175, Oct. 1986.
13. M. Dell'Amico and S. Martello, "The k-cardinality assignment problem," *Discrete Applied Mathematics*, vol. 76, no. 1, pp. 103–121, Jun. 1997.
14. A. Volgenant, "Solving the k-cardinality assignment problem by transformation," *European Journal of Operational Research*, vol. 157, no. 2, pp. 322–331, Sep. 2004.

Printed in the United States
By Bookmasters